Statistics in Geography

For Anna, Stephanie and Thomas

David Ebdon

Statistics in Geography

Second Edition
Revised with 17 Programs

BLACKWELL
Oxford UK & Cambridge USA

First published 1977
Reprinted 1978, 1981, 1983
Second edition 1985
Reprinted 1988 (with corrections), 1990, 1992 (twice), 1994, 1995,
1996, 1997, 1998, 1999, 2000

Blackwell Publishers Ltd
108 Cowley Road
Oxford OX4 1JF, UK

Blackwell Publishers Inc
350 Main Street
Malden, Massachusetts 02148, USA

British Library Cataloguing in Publication Data
A CIP catalogue record for this book is available from the British Library

Library of Congress Cataloging in Publication Data
Ebdon, David
Statistics in geography
Includes index.
1. Geography—statistical methods. I. Title.
G70.4.E23   1985   910'.01'5195   85–15660
ISBN   0–631–13688–6 (pbk)

Printed in Great Britain by T. J. International Limited, Padstow, Cornwall

This book is printed on acid-free paper

Contents

Preface to the Second Edition

Since the first publication of this book in 1977 there has been a minor revolution in educational technology — or more precisely a micro revolution. The introduction of micro computers into schools, colleges and universities means, or certainly should mean, that students can be much more extensive and perhaps more adventurous in their use of statistics. Freed from the tedium of manual calculation, they can concentrate more on the application and interpretation of statistical techniques.

The main new feature of this edition is a set of Basic computer programs that cover most of the statistical techniques described. The programs are written in an elementary subset of the Basic language which should make them simple to implement on most micro computers. The programs have been tested on a 32k BBCb micro. Because of the enormous number of different types of micro and the variety of implementations of Basic I can not guarantee that they will necessarily run unaltered on *any* micro, but only minor changes should be necessary.

It could be argued that the availability of computers makes it unnecessary for students to learn hand methods of calculation. I considered removing all the worked examples from this book. However, the micro revolution has not yet pervaded all parts of the education system in Britain, even less all parts of the world. In any case I believe that students can gain insight into the logic and interpretation of statistical methods by working through them by hand. The danger of an automated system for statistical analysis is that it might remove from the students the responsibility for thinking about what they are doing. Students, and for that matter professional geographers, who have access to powerful computerized statistical packages can easily be lured into the habit of trying every possible technique until they produce some results they like the look of!

Other changes in the book are relatively minor. A number of errors have been corrected, and I am grateful to various reviewers for pointing these out. I have also changed some of the examples to widen the appeal of the book to those readers who have the misfortune not to live in Nottingham, or even England!

Lecturers and others using the book may wish to comment on the content of the book, especially perhaps on the programs. If so, could you write to me care of the publishers at either their Oxford or New York offices (see page iv). Both I and they would be particularly interested to know whether or not you would welcome the availability of versions of the programs on disk.

David Ebdon
Nottingham, 1985

Preface to the First Edition

It is now generally accepted amongst professional geographers that a training in elementary statistics should form an essential part of any geography course. The majority of students accept this dictat with characteristic good humour. Nevertheless there is still a widely held feeling amongst the consumers that statistics practical classes are one of the more masochistic manifestations of geographical asceticism. This book is founded on the firmly held belief that statistics are vital to the development of geography as an academic, and more importantly a practical, discipline. If it helps in any way to overcome the initial consumer resistance in schools, colleges and university departments I shall be more than satisfied.

It is perhaps not difficult to understand why statistics are often only tolerated by the student population. Some of the concepts are indeed difficult to understand; it would be dishonest to claim otherwise. A rather more immediate objection concerns the peculiar jargon and the complex notation employed by statisticians and their geographical disciples. Fortunately, however, one of the main practical difficulties has largely been overcome in the past few years. The rapid fall in the price of electronic calculators means that at last the 'arithmetic barrier' has been well and truly broken. Furthermore, many students in schools, colleges and university geography departments now have access to mini-computers which can be programmed to solve statistical problems in a matter of seconds. Unfortunately this development brings with it the danger, already being realized in some university geography departments, that

students will use sophisticated statistical techniques at the drop of a computer card, without pausing to consider whether their application is valid or meaningful.

The aim of this book is to introduce the reader as painlessly as possible to some of the concepts and methods of statistical analysis. As with all things, there are various levels at which statistics can be understood. Some readers will be prepared simply to accept that a particular technique does what it is supposed to do, without needing to understand precisely how it does it. Provided he or she follows the instructions supplied with the technique, this type of reader will come to no great harm. For those who are more inquisitive some attempt is made in this book to explain the workings of statistics, almost as a piece of machinery. Wherever possible practical analogies are given rather than theoretical expositions. To this extent the book is intended to combine the roles of owner's handbook and workshop manual. To pursue the analogy a little further, it is also intended to act as a highway code, listing the rules that must be obeyed if valid statistical conclusions are to be reached.

The first chapter of the book explains a number of basic statistical concepts to which frequent reference is made in subsequent chapters. In the rest of the book selected statistical techniques are discussed. The choice of techniques for inclusion was governed by a number of considerations. First there are certain techniques which have a relatively long and respectable history in geographical research. These techniques are described partly because they are indeed useful aids to geographical analysis, and partly because they have been and still are frequently misused. A second group of techniques, mainly nonparametric, is included because these techniques deserve to be used more widely in geography. They are generally more robust and less easily mishandled than 'traditional' methods. Finally a few techniques are included because they represent a relatively recent trend in geographical statistics towards the study of inherently spatial patterns and relationships.

In all cases the choice has been firmly circumscribed by the need to include only those techniques which can be applied using manual, as opposed to computerized, methods of calculation. With this aim in mind the various equations given in the text have been carefully chosen or modified to achieve a balance between ease of understanding and speed of computation. No apologies are offered for the fact that mathematical notation has been kept as simple as possible. No knowledge of mathematics beyond 'O' level is assumed. Examples and exercises are also deliberately simple; many of them have been unashamedly fabricated. The problem with many real world examples is that they introduce computational complexities which are out of place in an elementary text such as this. It is worth adding that the exercises are seen as an integral part of the book. Consequently answers and workings for all the exercises are given in some detail in Appendix A.

Many people have helped, wittingly or unwittingly, in the preparation of this book. Thanks are due to John Cole for putting me up to it in the first place and for much helpful advice and criticism since then. Paul Mather also read parts of the manuscript. I am grateful for his comments and for the computer programs he pro-

vided for calculating critical values of t, χ^2 and F. I am equally grateful to Jim Feather for his help in the early planning stages, and for his patience in the face of undue prevarication towards the end. The following authors and publishers gave permission for me to use material in the preparation of Appendices B and C: E. G. Olds and the Institute of Mathematical Statistics (critical values of Spearman's rank correlation coefficient), Z. W. Birnbaum and the American Statistical Association (critical values of the Kolmogorov–Smirnov statistic), W. H. Kruskal, W. A. Wallis and the American Statistical Association (critical values of H for the Kruskal–Wallis test), D. Auble and the Institute of Educational Research at Indiana University (critical values of U for the Mann–Whitney test), Biometrika Trustees, E. S. Pearson and H. O. Hartley (critical values of the product-moment correlation coefficient), C. Eisenhart, F. S. Swed and the Institute of Mathematical Statistics (critical values for the runs test), and the Chemical Rubber Company (percentage points of the normal distribution). The co-operation of all these people is gratefully acknowledged.

The largest single group of helpers have been the students in the Geography Department at Nottingham University. Over the past three years their blank looks and awkward questions have made me try harder and think more carefully about the teaching of statistics. I hope I have not entirely failed them. Finally I should like to express my affectionate thanks to my wife, Gill, for putting up with my frantic nocturnal writing bouts, and to my daughter, Anna, for remaining cheerfully oblivious of the whole silly business.

David Ebdon
Nottingham, 1976

1 Statistical Concepts

1.1 The Uses of Statistics

It is difficult to give a satisfactory definition of statistics. In common usage the term often means 'facts and figures', for example health statistics or education statistics. In a technical sense statistics (or more correctly mathematical statistics) refers to a branch of applied mathematics concerned with the interpretation of numerical information. It is a science, or perhaps an art, which has been in existence only since the late 19th century, although the foundations of modern statistics were laid in the 17th and 18th centuries by mathematicians concerned with probability. As far as this book is concerned four main functions of statistics can be recognised: description, inference, significance testing and prediction.

Description

Geographers, like many other scientists both social and physical, are having to come to terms with an 'information explosion'. The amount of information, and with it the amount of numerical data, which is available to them is increasing at an accelerating rate. If the geographer is to make use of this mass of information he needs ways of summarizing large sets of data so that he has concise measures of their characteristics. *Descriptive statistics* can help to fulfil this aim, both for ordinary data sets and for spatial data (data relating to a two-dimensional surface or a three-dimensional space).

Inference

Most geographers have to deal with data obtained from samples, rather than with all the data they could possibly have about a particular situation. In such a case it would be convenient to assume that the sample is representative of the total set of data (the population) from which it has been drawn. *Inferential statistics* can enable the geographer, within certain strictly defined limits, to make statements about the characteristics of a population based only on data collected from a sample. Electoral opinion polls, for example, depend very much on inferential statistical techniques, although the spectacular inaccuracy of these polls should not be taken as typical of the results of inferential statistics!

Significance

One of the most powerful uses of statistics is in helping the geographer to decide whether an observed difference or relationship between two sets of sample data is significant. *Significance* (we should really call it 'statistical significance') means much the same here as it does in everyday usage. When we ask if something is significant (as radio and television interviewers are so fond of doing) what we are asking is 'does it signify anything?', 'does it really mean anything?'. So, in a sense, statistical significance is concerned with whether an observed difference, for example, between two samples, can be taken as signifying something else—that there is a difference within the population from which the samples were drawn, or alternatively that the difference appears in the samples merely as a result of chance in the sampling procedure. This will be explained in more detail in Section 1.4.

An important point to be made is that inferential statistics and tests of significance rely very heavily on the concept of probability. Statistics themselves cannot make judgements for us: they cannot make inferences about populations from samples, nor can they tell us that a relationship is significant. What they can do is provide reasonably objective information on which to base our own, and therefore unavoidably subjective, judgements. Statistics can tell us the probability (or likelihood) that, under certain specified conditions, a relationship is significant; the probability that, again under specified conditions, inferences made on the basis of samples are valid.

Prediction

A fourth major use of statistics is in helping us to make predictions, or indeed postdictions (statements about what may have happened at some past time). Completely accurate prediction is only possible where some completely *deterministic* process is in operation, by which we mean a process which will always produce identical results in identical circumstances. Acceleration due to gravity is an example of such a process. We know (or at least have to accept!) that in a vacuum acceleration due to gravity is always 9.75 m/sec/sec. Armed with this information we can predict with absolute certainty how far an object will fall in a given time, or what its speed will be at a particular instant in time during its fall, provided it is falling in a vacuum.

However, very few if any geographical processes are of a deterministic nature. They frequently behave in different ways at different times, and we can seldom be certain of the outcome of a process even under carefully controlled conditions. Nevertheless, provided the process is not entirely random (or haphazard) it may be possible to predict the outcome of a particular combination of circumstances within certain limits. For example, if a cliff face is being

eroded back at an average rate of so many metres per year it may be possible to predict the position of the cliff edge in five years time, to within a few metres.

Statistics can provide information about the regularity of the past and present workings of a process, and in this way can help to make possible some sort of probabilistic prediction. Ultimately, however, the prediction of a future event or of the future outcome of a process always involves an intuitive leap into the unknown. The fact that a process has been behaving in a certain way for the past days, years, or even centuries does not preclude the possibility that its behaviour will change completely in one second's time!

So far in this section a number of important concepts have been introduced in a rather superficial way. They will be explained more fully in the next three sections.

1.2 Data

'Data', which is a plural noun, means a body of information in numerical form. It is hard to see that there can be any data which are uniquely geographical, except perhaps data which concern the spatial characteristics of places and areas, for example, shape and pattern. However, the geographer is certainly concerned with numerical information from a wide range of sources. This book contains a set of data relating to the 50 largest towns in England and Wales (according to the 1971 census). These data are given in Appendix E and will be used in many of the exercises and examples which follow. A set of data in tabular form is often referred to as a data *matrix*. A matrix is equivalent to an array in computer programming (see Mather 1976). Statistical techniques are mainly applied to information which is expressed in numerical form, and although it may seem that numbers are just numbers there are some definitions that need to be given. They relate to terms that will be used repeatedly in this book.

Individuals and Variables

In statistics it is common to refer to the items about which information has been obtained as *individuals.* In the data matrix (Appendix E) the individuals are therefore the 50 towns to which the various pieces of information refer. Each column of figures in the matrix gives the values of a *variable*, a particular characteristic or property which varies from one individual to another. Bristol, Bolton and Nottingham are individuals in the data matrix. Population, rainfall and altitude are variables. This terminology will be used throughout this book, but alternatives may be met elsewhere, particularly *case* for individual and *variate* for variable.

Types of Number

As far as this book is concerned there are two types of number: integers and real numbers. An *integer* is simply a whole number, with no fractional or decimal part. Integers can be either positive, negative, or zero. A *real number* is any number which has a fractional or decimal part. So 47, −9, −27, 468, 0 and 14 are all integers, and 16.927, 0.03, −468.93, $6\frac{3}{4}$ are all real numbers.

Measurement Scales

The distinction between integers and real numbers will only occasionally need to be made in this book. The distinction between different scales of measurement will be important throughout, since particular statistical techniques can only be applied to data measured on a particular scale. Three main scales of measurement are recognised: nominal, ordinal and interval (including ratio).

NOMINAL

Measurement on a *nominal* scale consists of simply placing each individual into one of a number of different categories. In the data matrix (Appendix E) variable J is a nominal variable. Each of the towns has been allocated a number according to whether it has a local radio station. The figure 1 denotes that the town has a local radio station, 0 that it does not. These values are being used as labels for the two categories, and although they could be inter-

preted as the number of local radio stations each town has, this would be fallacious. London, Nottingham and several other towns have more than one local radio station. The same applies to variable M, which denotes whether or not a town has a university. London and Birmingham both have the value 1, although they both have more than one university. Other possible nominal categories into which each of the 50 towns could be put include ports, seaside towns, county towns, manufacturing towns and so on.

ORDINAL

Measurement on an *ordinal* scale involves putting individuals into an order, ranking them according to some criterion. Some data are inherently of an ordinal nature, for example, data on preferences. In the rapidly expanding field of perception studies, geographers are often asking people to rank towns, parts of the country, or photographs of different types of scenery in order of preference. In other cases it may be decided to put data into ordinal form, for example variable B in the data matrix. This variable has been created 'artificially' by ranking all the towns according to their population totals, from the largest to the smallest. This same process could be carried out for many of the other variables in the matrix. Ordinal data is one step up from nominal data in the sense that it enables us to say whether one individual comes before or after another along a scale.

INTERVAL

Measurement on an *interval* scale consists of allocating a number to an individual to indicate its precise position along a continuous scale. For example, temperature can be measured in degrees centigrade to any desired number of decimal places, depending only on the precision of the thermometer. Interval data is another step up the heirarchy of measurement scales, in the sense that it enables us to say how much further along a scale one individual is than another. The rainfall variable tells us by how much one town is wetter than another. If the rainfall figures were to be ranked, they would merely tell us that one town was wetter than another, but not by how much.

Measurements made on a *ratio* scale have all the characteristics of interval measurements, with the added feature that the ratio of any two values on a ratio scale is independent of the unit of measurement. For example, if one statistics book weighs 1 lb. and another 2 lbs., the ratio of their weights is $1/2 = 0.5$. If metric units were used, their weights would be 453 grammes and 906 grammes, but the ratio of their weights would still be 0.5 (453/906). However, on a non ratio scale, such as temperature, this is not the case. At noon on two successive days the temperature is 40°F and 68°F. The ratio of these two temperatures is $40/68 = 0.59$. The same two temperatures in centigrade would be 10°C and 20°C, but the ratio between them would now be 0.50 (10/20).

The important distinction is between nominal, ordinal and interval scales. Ratio data can be dealt with statistically in just the same way as interval data. A ratio variable is also an interval variable, but an interval variable is not necessarily ratio. In the rest of this book the term interval should be taken as meaning interval and/or ratio.

The distinction between nominal, ordinal and interval data can be reinforced by considering how one form can be converted into another. The figures for annual rainfall (variable K in Appendix E) are presented in interval form. They enable us to say, to the nearest millimetre, how much more rain one town receives than another. These data could be converted into ordinal form by ranking the values from 1 to 50, from the highest to the lowest. In doing this some information would be lost. It would still be possible to say, for example, that Bolton is wetter than Bristol, but not just how much wetter it is. A further loss of information would result from conversion to a nominal scale. Each town could, for example, be categorized as wet or dry, according to whether its rainfall total is above or below the average for all the towns. Bristol would now be classed alongside Bolton as a wet town.

It may seem a rather perverse exercise to discard information which has probably cost someone a lot of time and money to collect. However, it can be an important means of ensuring that statistical tests can subsequently be applied to the data, as will be seen in later chapters.

Errors and Precision

Statistics can give a beguiling impression of precision. Answers can be given to any number of decimal places. However, geographical measurements, particularly in the currently fashionable areas of human geography such as behavioural studies, are subject to error. In some cases the geographer may have a good idea what sort of errors are involved. A theodolite can be accurate to a very small part of a degree. The Office of Population Censuses and Surveys publishes detailed estimates of the errors contained in census volumes. In other situations, such as the small scale surveys often carried out by human geographers, there may be no way of knowing the magnitude of measurement errors involved. What can be done about this?

Certainly there is no need to despair. Statistical techniques may still be applicable, but geographers should beware of *spurious precision*. Giving statistical measures to six decimal places, when the original data are only accurate to the nearest whole number, certainly involves an element of self-deception. It may be, for example, that a particular variable has been measured on an interval scale with a high degree of error. Results of statistical tests applied to such data are unlikely to be very reliable, no matter how accurately the calculations have been carried out. However, that variable may still be usable if converted into ordinal form, provided the original measurements accurately record the fact that one value is greater than or less than another.

Presentation of Data

Quite a lot can be learnt (or guessed) about a set of data by displaying it in some appropriate visual form. This is a very necessary prelude to statistical analysis, since it may prevent the geographer from ignoring some of the major peculiarities or shortcomings of his data. It may also bring to light relationships within the data that cannot be seen just by looking at a mass of figures. Two simple methods of presentation will be briefly discussed here: histograms and scatter graphs.

HISTOGRAMS

A *histogram* (or bar diagram) is a convenient means of showing the distribution of values in a set of data. Figure 1.1 is a histogram of the rainfall data (variable K in Appendix E) for the 50 towns. The range of rainfall totals has been

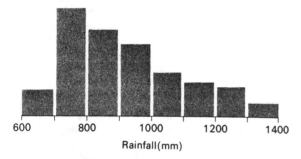

Figure 1.1 Histogram of rainfall totals

divided up into a number of classes. For each class the size of the bar is proportional to the number of towns in that class. This number is often referred to as the *frequency* of the class, and so a histogram is a visual representation of the *frequency distribution* of a variable.

SCATTER GRAPHS

A *scatter graph* is just a plot of the values of one variable against values of another. Figure 1.2 is a scatter graph of altitude against rainfall for the 50 towns. From the scatter graph no very clear relationship between altitude and rainfall is apparent. It is not possible to say that rainfall either increases or decreases systematically with altitude.

1.3 Probability

Probability is a fundamental concept in statistics. All statistical tests involve the calculation of probabilities, either directly or indirectly. Statistics is concerned not with certainties but

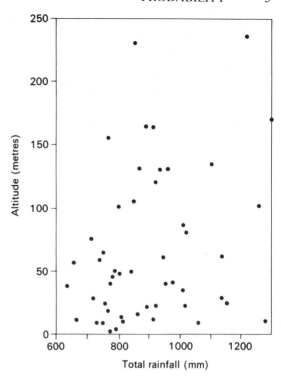

Figure 1.2 Scatter graph of rainfall and altitude

with probabilities. Statistical hypotheses are never said to be true or false. Rather, the **probability** that they are true or false is stated. The first aim of this section is to show what probability is and how it can be calculated. Secondly some simple rules of probability will be given to show how probabilities can be combined. Thirdly the idea of a probability distribution will be introduced. In particular two important theoretical probability distributions will be discussed.

Calculating Probabilities

The concept of probability can best be demonstrated by example. If a coin is tossed the probability that it will come down 'heads' is $\frac{1}{2}$. There are two possible outcomes, 'heads' or 'tails'. Only one of these outcomes is 'successful', in the sense that it produces the desired result (in this case a 'head'). The probability of getting a six when a die is thrown in 1/6. There are six possible outcomes, 0, 1, 2, 3, 4, 5 and 6, but only one of these is 'successful'. If two coins are tossed, the probability that they will both come down 'tails' is 1/4. There are four possible outcomes: head/head, head/tail, tail/head, tail/tail, only one of which is 'successful'. Note that in each of these three examples the outcomes are all equally likely (probable). If they were not, then games of chance based on coin tossing and the throwing of dice would be rather unfair.

One definition of probability is therefore: *the ratio of the number of successful outcomes to the number of all possible outcomes*. The probabilities given above are theoretical probabilities; they can be calculated in advance rather than found by experiment. This kind of probability is *a priori* probability, sometimes Anglicized to 'prior probability'. Probabilities which can only be found by experiment are known as *a posteriori* probabilities, or (a rather unfortunate piece of transliteration) 'posterior probabilities'. The concept of *a posteriori* probability can also be demonstrated by example.

A lecturer gives 20 lectures in a term and overruns the allotted time on 5 occasions. One quarter of his lectures overran, or in other words the *a posteriori* probability that any **one** of his lectures would overrun was 1/4. This could not be known at the time; it can only be calculated afterwards. Assuming the lecturer does not undergo a personality change during the vacation, it can be estimated that the probability that any one of his lectures will overrun in the future is 1/4.

Probabilities can be expressed in various forms: as fractions, percentages or decimal fractions. For example, one chance in five gives a probability of 1/5, or 20%, or 0.2. Statisticians tend to calculate probabilities as decimal fractions and this is the practice that will mostly be followed in this book. However probabilities are expressed, three important points should be borne in mind:

(1) If an outcome is certain to be 'successful' it has a probability of 1.

(2) If an outcome is certain not to be 'successful' it has a probability of 0.

(3) The individual probabilities of all possible mutually exclusive outcomes must add up to 1. For example, if a die is thrown the probability of each individual outcome (1, 2, 3, 4, 5 or 6) is

1/6. These six individual probabilities added together come to 1.

Combining Probabilities

Not all probabilities are as simple to calculate as the ones described in the previous section. Sometimes it is necessary to combine the probabilities of two or more events or two or more outcomes. The term 'event' is used by statisticians to refer to something like tossing a coin, or throwing a die, or picking a lottery ticket from a hat. Each 'event' has an 'outcome' which may or may not be 'successful'. Two rules govern the way in which probabilities can be combined.

MULTIPLICATION RULE

Two dice are thrown. What is the probability that the result will be a six on one and a six on the other? Two events are being combined in such a way that a successful overall outcome (two sixes) requires a particular type of outcome (a six) from both individual events (throws of a die). In this situation the individual probabilities must be multiplied to give the overall probability. Each individual probability (that a die will give a six) is 1/6, so the overall probability is $1/6 \times 1/6 = 1/36$.

ADDITION RULE

A dice is thrown. What is the probability that the result will be a one or a two or a three? Note that these are three mutually exclusive outcomes. The probability of each individual successful outcome is 1/6. There is a probability of 1/6 of getting a one, and the same probability applies to a two, or a three. To find the overall probability in this case the individual probabilities must be added. The overall probability of getting a one or a two or a three is therefore $1/6 + 1/6 + 1/6 = 1/2$.

In general the word 'and' implies multiplication of probabilities, and the word 'or' implies addition. If there is any doubt about which rule to apply, it is usually possible to test all the possible outcomes, count up the successful ones and divide by the total number of outcomes to give the probability. In the case of the addition example above, the total number of outcomes is six, and the number of successful outcomes is three. The probability is therefore $3/6 = 1/2$.

Two further definitions are relevant to the discussion of combined probabilities. If two events occur and the outcome of one can in no way influence the outcome of the other, the events are said to be *independent*. The two events also have independent probabilities. For example, if a coin is tossed a number of times each outcome (heads or tails) is independent of all the other outcomes in the sequence. The result of one toss cannot influence the result of the next. The probability of getting a head is the same $(1/2)$ every time. Even if 10 heads come up in a row the probability that the next toss will produce a head is still $1/2$.

If the result of one event does influence another, the second is said to be *dependent* on the first. For example, suppose there are ten names in a hat, one of which is yours. When the first draw is made the probability that your name will be picked out is $1/10$. If your name is not picked out the first time, the probability that it will come up at the next draw is $1/9$ (there are now only 9 names left in the hat). This is on the assumption that the names are not put back into the hat after each draw, i.e. that the draw is being made *without replacement*. The probability will go on reducing in this way as the number of names left in the hat diminishes. As soon as your name has been picked, the probability that it will be picked the next time is zero.

Probability Distributions

In Section 1.2 the concept of a frequency distribution was introduced. It is quite a simple task to convert a frequency distribution into a *probability distribution* by replacing the absolute frequency for each class by its proportional frequency. Table 1.1 gives the frequency distribution of the children in care variable (from Appendix E). It also shows how a simple frequency distribution can be converted into a probability distribution. Column a contains the definitions of the classes, and column b gives the frequency for each class. The total number of items in the frequency distribution is 49 (the sum of all the individual frequencies). The probabilities in column c are calculated by dividing each frequency by the sum of the frequencies. The first frequency is 3, so the corresponding probability is $3/49 = 0.0612$. The second frequency is 7, so the probability is $7/49 = 0.1429$. The rest of the probabilities are calculated in the same way. Note that the sum of all the individual probabilities is 1.0. Only 49 individuals are included in the frequency distribution because data are not available for one town (Basildon).

THE NORMAL DISTRIBUTION

The normal distribution is the most important probability distribution in statistics. It forms the basis of a large group of statistical tests known

Table 1.1 Frequency and Probability Distribution of Children in Care

a class	b frequency	c probability	
2– 3.99	3	0.0612	(3/49)
4– 5.99	7	0.1429	(1/49)
6– 7.99	18	0.3673	(18/49)
8– 9.99	10	0.2041	(10/49)
10–11.99	9	0.1837	(9/49)
12–13.99	1	0.0204	(1/49)
14–15.99	1	0.0204	(1/49)
	49	1.0000	

as *parametric* techniques. Several of these techniques will be discussed in later chapters of this book. The mathematics of the normal distribution are quite complicated and will not be discussed here. However, it is important to understand what the normal distribution looks like and how it is possible to compare an observed distribution of data values with a theoretical normal distribution.

Figure 1.3 shows a normal probability distribution. It can be seen that it is symmetrical and bell-shaped. The horizontal scale along the

possible to calculate the probability that a value of that variable for a randomly chosen individual will lie between any specified limits. This probability can also be used to give the proportion of all the values of the variable that should be between the specified limits. *Fitting a normal distribution* to a set of data is the expression used to describe the process of comparing the observed data distribution with a theoretical normal distribution. The following example describes how a normal distribution is fitted to a set of data.

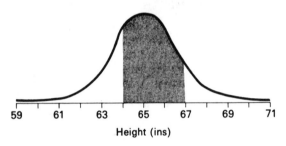

Figure 1.3 Normal distribution of heights

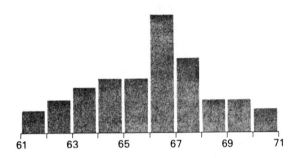

Figure 1.4 Sample distribution of heights

bottom of the distribution represents values of a variable, in this case height of female geography students. The mid point on this scale is the *mean* ('average') of the values, in this case the mean height. The mean will be discussed in detail in Section 2.1. For the moment it is sufficient to know that it is the usual measure of 'average', calculated by adding together all the values and dividing by the number of values.

The total area under the curve represents a probability of 1.0. The area under the curve between any two limits on the horizontal scale is proportional to the probability of occurrence of a value of the variable (height) which is somewhere between those limits. For example, the shaded area in Figure 1.3 gives the probability that a female geography student chosen at random will be between 5' 4" and 5' 7" tall. For the sake of argument, it will be assumed that this probability is 0.55. This probability implies that 55% of all female geography students should be between 5' 4" and 5' 7" tall.

In summary, if it can be assumed that a particular variable is normally distributed, it is

Figure 1.4 is a histogram of the heights of a sample of female geography students. Table 1.2 gives the frequency distribution of these data, and the probability distribution (columns b and c). The aim is to find the corresponding *normal probabilities*, in other words, the probabilities that would apply if height of female geography students really were a normally distributed variable. The steps involved in fitting a normal distribution will now be described.

(1) Convert the class boundaries into units of *standard deviation*. Standard deviation is a measure of the 'spread' of values in a frequency distribution. It is discussed in detail in Section 2.2. For the moment it is not necessary to know how it is calculated, only to accept that a large standard deviation implies that there is a wide spread of values, whereas a small standard deviation implies that many of the values are close together in size.

To convert a value into a unit of standard deviation (or *standard deviate*) it is necessary to

Table 1.2 Fitting a Normal Distribution to a Set of Height Data

a	b	c	d	e	f
			class	standard	normal
class	frequency	probability	boundary	deviate	probability
61"+	2	0.044	61"	−2.00	0.035
62"+	3	0.067	62"	−1.57	0.071
63"+	4	0.089	63"	−1.13	0.113
64"+	5	0.111	64"	−0.70	0.155
65"+	5	0.111	65"	−0.26	0.170
66"+	11	0.244	66"	0.17	0.162
67"+	7	0.156	67"	0.61	0.122
68"+	3	0.067	68"	1.04	0.080
69"+	3	0.067	69"	1.48	0.041
70"+	2	0.044	70"	1.91	0.018
			71"	2.35	
	45	1.000			0.967

subtract the mean from it and then divide by the standard deviation:

$$\text{Standard deviate} = \frac{\text{value} - \text{mean}}{\text{standard deviation}} \quad (1.1)$$

Column e of Table 1.2 gives the class boundary values of the frequency distribution converted into standard deviates. The mean height of the sample of students is 65.6", and the standard deviation of their heights is 2.3". The first class boundary is 61", so this becomes:

$$\frac{61 - 65.6}{2.3} = -2.0 \text{ standard deviates}$$

The other class boundaries have been converted in the same way. This process of *standardization* is necessary so that the normal distribution can be fitted without needing to worry about the original units of measurement, in this case inches.

(2) Consult a table of the normal distribution (Appendix B1) to find the normal probability that a value will occur between the limits set by the class boundaries. The total area under the normal curve is considered to be equal to 1000 for the purpose of the table. Each number in the body of the table gives the area **to the left of** a particular number of units of standard deviation, specified by the row and column labels. For example, the row labelled −1.2 and the column labelled 0.03 refers to −1.23 units of standard deviation. The number at the intersection of this row and column is 109. This is the area under the normal curve **to the left of** −1.23 units of standard deviation. Since the total area is 1000, the probability that a value in a normal distribution will be **less than** −1.23 standard deviates, is therefore, 109/1000, or 0.109.

The probability that a value in a normal distribution will lie **between** two specified standard deviates, can be found by subtraction. For example, the first two class boundaries in Table 1.2 are −2.00 and −1.57 standard deviates. From Appendix B1 the probability that a value in a normal distribution will be less than −2.00 is found to be 0.0228 (22.8/1000). The normal probability of a value less than −1.57 is 0.058 (58/1000). The normal probability of a value between −2.00 and −1.57 is therefore, 0.0352 (0.058−0.0228). The normal probabilities for all the other classes in the frequency distribution of heights are calculated in the same way. They are given in column f of table 1.2. It will be seen that the normal probabilities do not (quite) add up to 1.0. This is because in the normal distribution of heights it is possible to find values outside the observed range. In other words, according to the normal distribution there is a chance of finding female geography students who are under 5′ 1″ tall or over 5′ 11″ tall, even though no people this short or this tall occur in the sample. The observed distribution is therefore being compared with only part of the theoretical normal distribution.

The normal probabilities can now be compared with the observed (*a posteriori*) probabilities to see how well the observed distribution corresponds to a normal distribution. Statistical tests for measuring this correspondence are known as goodness-of-fit tests. One such test, the Kolmogorov–Smirnov test is described in Section 4.1.

In passing, it might be noted that a standard deviate is often denoted by z, and even referred to sometimes as a *z value*. Furthermore, a standard deviate relating to a variable which does have a normal distribution (not just a variable which is being compared with a normal distribution) is known as a *standard normal deviate*. Standard normal deviates will be met again in connection with certain statistical tests relating to spatial situations (Chapter 7).

The reader will find it helpful to memorize the following relationships between units of standard deviation and normal probabilities:

(1) The probability that a value in a normal distribution will be between the limits set by one unit of standard deviation on either side of the mean is 0.682.

(2) The probability that a value in a normal distribution will be between the limits set by two units of standard deviation on either side of the mean is 0.954.

(3) The probability that a value in a normal distribution will be between the limits set by three units of standard deviation on either side of the mean is 0.997.

These probabilities will be met several times in later chapters.

Exercise 1.1

Fit a normal distribution to variable E (children in care) from Appendix E. The data have already been put into a frequency distribution in Table 1.1. The mean is 7.896, and the standard deviation is 2.495.

THE POISSON DISTRIBUTION

The Poisson distribution is one which occurs quite frequently in geographical studies. It is concerned with random occurrences. The classic example concerns deaths in the Prussian Army as a result of horse-kicks. It is of no geographical significance whatsoever, but it may be sufficiently whimsical to stick in the reader's memory. Table 1.3 shows number of years in which there were 0, 1, 2, 3, and 4 deaths from horse-kicks. These data are derived from annual records of each of 10 army corps kept for a period of 20 years, a total of 200 annual records. The data set could be thought of as relating to a total of 200 'corps-years' (by analogy with 'man-hours' or 'man-days'). Table 1.3 also gives the observed (*a posteriori*) probability of a particular number of deaths per 'corps-year'.

Table 1.3 Frequency Distribution of Deaths from Horse-Kicks

Number of deaths	Number of annual reports of this many deaths	Probability
0	109	0.545
1	65	0.325
2	22	0.110
3	3	0.015
4	1	0.005
	200	1.000

Deaths from horse-kicks can be expected to occur at random through time. There should be no way of predicting when a fatal horse-kick will be received. It is not possible to say that in a particular year there will be so many deaths from horse-kicks. However, it is possible to calculate the probability of a particular number of deaths in a year. This is what the Poisson distribution does. The calculation of the Poisson probabilities for these data will be briefly demonstrated.

The probabilities of 0, 1, 2, 3 and 4 deaths are given by the following series of expressions:

$$\frac{1}{e^z}, \quad \frac{z}{e^z}, \quad \frac{z^2}{2!e^z}, \quad \frac{z^3}{3!e^z}, \quad \frac{z^4}{4!e^z} \qquad (1.2)$$

where e is a mathematical constant which has the approximate value 2.7183 (like π, e cannot be expressed exactly as a decimal), and z is the average number of deaths (per corps-year). In the present example, there is a total of 122 deaths in 200 corps-years, giving an average of 0.61 (122/200). The symbol ! used after a number means *factorial*. Four factorial (4!), for example, is $4 \times 3 \times 2 \times 1 = 24$, and three factorial is $3 \times 2 \times 1 = 6$. The symbol e^z means 'e to the power z'. It is possible to calculate e^z (in this case $e^{0.61}$) using logarithms or on certain electronic calculators. For convenience, a table of values of e^z (calculated by computer) is given in Appendix B2. From the table it will be seen that $e^{0.61}$ is 1.8404. Each term of the Poisson series (1.2) can now be calculated:

0 deaths: $\dfrac{1}{e^z} = \dfrac{1}{1.8404}$ $= 0.543$

1 death: $\dfrac{z}{e^z} = \dfrac{0.61}{1.8404}$ $= 0.331$

2 deaths: $\dfrac{z^2}{2!e^z} = \dfrac{0.61^2}{2 \times 1.8404} = \dfrac{0.3721}{3.6808} = 0.101$

3 deaths: $\dfrac{z^3}{3!e^z} = \dfrac{0.61^3}{6 \times 1.8404} = \dfrac{0.2270}{11.0424} = 0.021$

4 deaths: $\dfrac{z^4}{4!e^z} = \dfrac{0.61^4}{24 \times 1.8404} = \dfrac{0.1385}{44.1696} = 0.003$

The more mathematically inclined reader may notice that there is a quick way of calculating the successive terms of the Poisson series. The second term is the first term multiplied by z and divided by 1. The third term is the second term multiplied by z and divided by 2. The fourth term is the third term multiplied by z and divided by 3. In other words, it is only necessary to work out the first term fully. The remaining terms can each be derived from the preceding one by simple multiplication and division. The probabilities given above can be compared with the observed probabilities given in Table 1.3. It will be seen that the correspondence is quite good. A goodness-of-fit test (Section 4.1) could be applied to measure the correspondence precisely.

The Poisson probabilities for deaths from horse-kicks, demonstrate a major characteristic of the Poisson distribution. There is a high probability of relatively small numbers of occurrences (deaths per year), and a low probability of a large number of occurrences. This is illustrated in Figure 1.5, which shows a Poisson probability distribution fitted to the horse-kicks data. The area of each bar is proportional to the

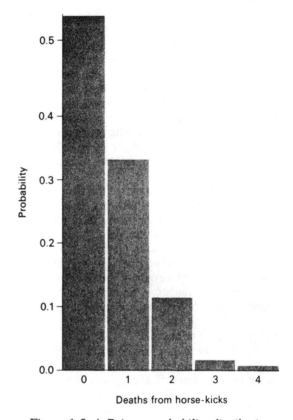

Figure 1.5 A Poisson probability distribution

probability of that number of deaths per year. It should also be noted that the Poisson probabilities given above do not (quite) add up to 1.0. This is because they do not cover all the theoretically possible outcomes. The Poisson series is an infinite series; it can be extended indefinitely. Only the first five terms are given in Equation 1.2. From the series it is possible to calculate the probability of 5, 6, 10, or even 100 deaths per year, although these probabilities would be very small.

Exercise 1.2

What are the probabilities of a) 5, b) 6, c) 10 deaths from horse-kicks per year, based on the data given above?

The Poisson distribution is applicable to very many geographical situations, since it relates to random occurrences in either time or space. Natural disasters, such as freak storms, are random events in time. It is necessary to say 'freak' storms, since storms of normal severity may be known to be more common at certain times of year. The Poisson distribution can be used to calculate the probability that there will be any specified number of freak storms in a year, based on the average number of freak storms per year. The average, like the average number of deaths from horse-kicks, can be calculated from past evidence, in this case from meteorological records.

For a spatial application of the Poisson distribution, consider the following example. As part of a recreation survey an aerial photograph of a moorland area is taken on a Bank Holiday weekend. A grid of four hundred 100 metre squares is superimposed over the photograph to enable a study to be made of the distribution of campers (their tents can be picked out on the photograph). A frequency distribution is produced, showing the number of 100 metre squares which contain 0, 1, 2, 3, up to 6 campers. This frequency distribution is given in Table 1.4. If campers can be considered to be random occurrences in space, the probability of finding any specified number in a 100 metre square can be calculated from the Poisson series.

Table 1.4 Data for Exercise 1.3

No. of campers	No. of squares
0	278
1	92
2	25
3	4
4	0
5	0
6	1

In this case the object of fitting a Poisson distribution to the data would be to see whether the phenomenon in question (distribution of campers) was or was not random. Note that the distribution of campers is a **frequency** distribution, not a **spatial** distribution. Nothing is being said here about the way in which the campers are arranged in space, which is what geographers often understand by the term 'distribution'. For example, it is not possible to say from the evidence in Table 1.4, whether 'crowded' 100 metre squares are clustered together in one part of the area, or whether they are widely dispersed. The topic of spatial distributions will not be dealt with until Chapter 7.

Exercise 1.3

Fit a Poisson distribution to the frequency distribution given in Table 1.4. Remember that it will first be necessary to convert the frequency distribution into a probability distribution. Be particularly careful to calculate z correctly (check with the answer in Appendix A before going on to the rest of the calculations).

There are numerous probability distributions which are applied to geographical situations. Only two have been discussed in this chapter, but they should at least serve to illustrate the way in which all probability distributions 'work'. The normal distribution must be understood because it forms the basis of so much of statistical theory. In later chapters, very many references will be made to the normal distribution. The Poisson distribution occurs quite frequently in geography, because of its association with randomness. By comparing an observed distribution of frequencies with a Poisson distribution, it is possible to decide whether or not, the frequencies are the result of a random process. The implication here is that, if the frequency distribution is **not** random, then it is worth looking at the situation more closely, in the hope of establishing cause and effect.

1.4 Significance

In the preceding sections, some of the basic concepts of statistics have been introduced.

Later chapters of this book are concerned with describing and explaining specific statistical techniques, each of which is applicable to a particular type of problem. Although these techniques are different in detail, they share a common approach. It is this approach, the philosophy of statistical significance testing which will be discussed in this chapter. The application of any statistical test involves a number of steps, each of which must be correctly carried out if the results of the test are to be valid. In the sections which follow, these steps will be dealt with in turn.

Hypothesis

A geographer studying a problem, may be struck by an apparent relationship in the data. There may appear to be a difference in vegetation between two areas of different soil type, or an increase in rainfall with altitude. From the results of questionnaire surveys, it may seem that different social classes have different shopping patterns, or that different age-groups have different perceptions of their home town. The words 'apparent' and 'seem' are justified here, since human beings are notoriously subjective and often overeager to see what they want to see. If the geographer wants his research to stand up to scientific scrutiny, he must adopt an approach which can be seen to be as objective as possible.

Part of this approach involves stating very clearly the hypothesis that is to be tested. This hypothesis is known as the *null hypothesis*. It is that the apparent relationship found in the sample data is **not** representative of a relationship in the population from which the data have come. In other words the null hypothesis is the opposite of what the researcher would like to believe. An *alternative hypothesis* is also set up. This is that the apparent relationship in the sample data does accurately reflect a relationship in the population. Setting up a null hypothesis may seem a rather cumbersome procedure, but it ensures that a suitably impartial attitude is adopted. In putting forward any case, it is fairer to the opposition to start out with the assumption that one is wrong, and then try to prove oneself right! An analogy can be made with the British judicial system, under which a suspect is, at least in theory, assumed innocent until proved guilty. An example may help to show how the two hypotheses are set up.

An intrepid biogeographer has spent several months in the depths of the New Forest, measuring numerous environmental variables, and making an inventory of vegetation cover at each of a number of sample points throughout the woodlands. After studying the mass of data it becomes apparent that a particular species of plant, say bracken, tends to occur in areas where there are steep slopes. The biographer feels that there may be a relationship between slope angle and the occurrence of bracken in the woodlands as a whole. The null hypothesis set up in this case would be that there is no relationship in the population between slope angle and occurrence of bracken. The relationship apparent in the sample data is due to chance in the sampling process. In other words, if it were possible to visit every point in the woodlands, measuring slope angle and counting the number of bracken plants, no relationship between the two variables would be found. The implication is that the sample of points just happens to be one in which a relationship is apparent, but that the sample is not representative of the woodlands as a whole.

Figure 1.6 shows a series of scatter graphs of two variables. Each dot represents a point at which a pair of measurements of two variables, such as slope angle and abundance of bracken, have been made. Figure 1.6a represents the population situation. It shows the values of the variables at all the points where measurements could possibly be made. This graph shows that there is no clear relationship between the two variables in the population; they do not vary together in any systematic way. The three other graphs in Figure 1.6 relate to results obtained from three different sample studies carried out on the population. A sample result like that shown in Figure 1.6b, would suggest that there is a *direct* relationship between the two variables; as one increases so does the other. Figure 1.6c shows a sample result in which there is an *inverse* relationship between the two variables; as one variable increases the other decreases. The result shown in Figure 1.6d suggests that there is no relationship between the two variables.

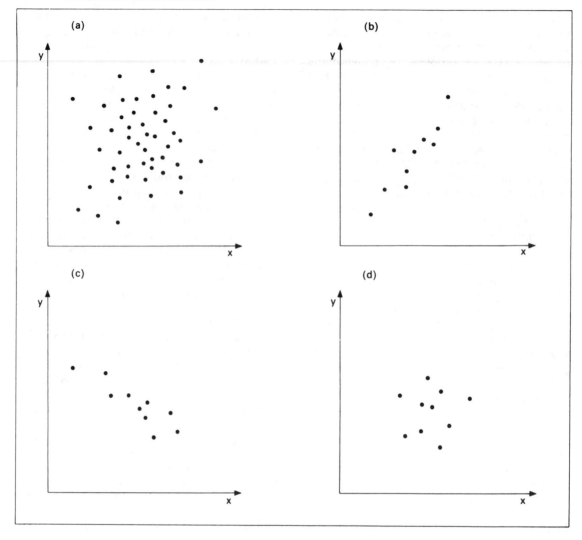

Figure 1.6 Population and sample relationships

Any of these three results could be obtained by taking samples from the same population. However, only one of them (Figure 1.6d) gives an accurate picture of the situation in the population as a whole. In this hypothetical case the situation in the population is known. The question is, without access to the population, how can it be known whether a set of sample results are an accurate reflection of the total situation? In fact, it can never be **known**, but provided certain assumptions are made, it is possible to make a reasonable estimate of the **probability**

that the sample results are representative of the population.

It is worth repeating however that the statistician's approach is to start from the assumption that the sample results are **not** representative of the population. This is his null hypothesis, conventionally denoted by the symbol H_0. There will also be an alternative hypothesis, denoted by H_1, that the sample outcome accurately reflects the situation in the population as a whole. The aim of an inferential statistical test is to calculate the probability that the null hypothesis is true. If this probability is acceptably low, the null hypothesis can be rejected in favour of the alternative hypothesis. The sample results can be said to be *significant*.

Assumptions

Statistical tests can conveniently be divided into two major classes: parametric and nonparametric techniques. *Parametric*, or classical, techniques make certain assumptions about the distribution of values in the population from which samples are taken. *Nonparametric*, or distribution-free, methods do not involve assumptions about the population. All the parametric techniques discussed in this book contain the assumption that sample values have a normal distribution. It is generally argued that a parametric test, used in a situation where its assumptions are justified, is more powerful than an equivalent nonparametric method. However, in many situations the geographer has no justification for making any assumptions about the population from which samples have been taken. Whether a nonparametric or a parametric test is used, one important assumption is **always** made in inferential statistics. This assumption is that the sample data come from *random* samples. The precise characteristics of a random sample will be discussed in Section 3.1. For the moment, random can be taken to mean unbiased. The application of inferential statistics to sample which are biased (not random) is invalid and can produce erroneous results. Equally, inferential statistics are not applicable to data relating to total populations rather than just samples taken from them. It is nonsense to talk about the probability that a sample situation accurately represents the population if it is the whole population that is being studied.

Other limitations on statistical tests concern the measurement scale (nominal, ordinal, or interval) of the sample values. All the parametric tests discussed in this book require variables to be measured on an interval (or ratio) scale. Nonparametric tests are less demanding and can be applied to ordinal or nominal measurements. The choice of a particular test thus depends not only on the nature of the problem, but also on whether any assumptions can be made about the distribution of values in the population. The scale of the variable measurements also needs to be considered. The assumptions and limitations of each test dealt with in this book are clearly stated in the appropriate section.

Test Statistic

Each of the statistical techniques discussed in this book involves the calculation of a single value called the *test statistic*. The test statistic has two functions. First it provides a description of the sample situation. Secondly it enables the significance of a relationship apparent in sample data to be evaluated.

Calculating the test statistic is often a laborious task, unless a calculator or computer is available. However, although it must be done accurately, it is possibly the least important step in the whole statistical procedure. Far more important is an understanding of the limitations and assumptions of the test, and an appreciation of its implications for the research in question.

In its descriptive function the test statistic gives an indication of the strength of a relationship between samples, or between variables within samples. Thus a low value of a particular test statistic may indicate a small degree of difference between two samples, or a high value of another may indicate a strong relationship between two variables. The second function of the test statistic depends on a knowledge of its sampling distribution.

Sampling Distribution

The *sampling distribution* of a test statistic is the probability distribution of values of that statistic in a situation in which the null hypothesis is true, i.e. in statistical jargon, *under the null hypothesis*. This can best be illustrated by example. Student's t is a test which measures the degree of difference between two samples. It is discussed in detail in Section 4.4. The null hypothesis of Student's t is that the two samples have come from a common, normally distributed population. Any difference between the samples does not therefore, reflect a 'real' difference. The alternative hypothesis is that the samples come from two different populations. The range of possible values of t is between zero and infinity. A low value of t suggests little difference between the samples. A high value is indicative of a large difference, and therefore provides more justification for rejecting the null hypothesis. However, not all values of t are equally likely to occur under the null hypothesis. It is to be expected that most pairs of samples taken at random from a common, normally distributed population (the null hypothesis situation) will be quite similar. They will produce low values of t. Only a few pairs of samples will be very different and give rise to large values of t.

Suppose a large number of pairs of samples were to be taken at random from the population, and the t test applied to each pair, a frequency distribution (histogram) of the t values resulting from each pair could be produced. This is the sampling distribution of t, and Figure 1.7 shows what it looks like. Values of t increase from left to right along the bottom axis, and the height of the curve is proportional to the number of times particular values can be expected to occur.

The distribution shown in Figure 1.7 can also be thought of as a probability distribution. The total area under the curve is equal to 1.0 (the probability of getting a value of t somewhere between zero and infinity). From this distribution it is possible to calculate the probability that a sample study will produce a value of t between any specified limits. For example, the shaded area in the *tail* of the distribution represents the probability of getting a value of t

greater than t_c. The benefit of being able to calculate this probability will be seen later.

Significance Levels

The probability that the null hypothesis is correct is referred to as the *significance level*. Earlier it was said that the null hypothesis can be rejected if this probability is acceptably low. Just what is meant by 'acceptably low' is, however, open to discussion. Traditionally statisticians have tended to be extremely cautious, and to tolerate only very low probabilities of being wrong in rejecting the null hypothesis, probabilities as low as 0.01 or 0.001.

Conservative significance levels such as these are certainly required in areas such as medicine or civil engineering, where lives may depend on the correct interpretation of statistical tests. It may seem heretical to suggest this, but geographical research is seldom so important (in a 'life-or-death' sense) that such conservative significance levels are necessary. Significance levels of 0.05, or even 0.1, are possibly adequate in many geographical applications. However, there is certainly never any harm in erring on the side of caution.

The important point is that, in order to be objective, it is essential to decide on the significance level **before** carrying out the test. Once the significance level has been chosen, the next step in the procedure is to find the critical value of the test statistic.

Critical Values

Without referring to it as such, the *critical value* of a test statistic has already been defined. If the shaded area in the tail of the sampling distribution (figure 1.7) represents a probability of 0.05 (5% of the total area under the curve), then t_c is the critical value of t at the 0.05 significance level. This means that there is a probability of 0.05 of obtaining under the null hypothesis a value of t between t_c and infinity. Suppose the t test is applied to a sample study using the 0.05 significance level, and the calculated value of t is greater than t_c. This means that there is a probability of not more than 0.05 (5 chances in 100) that the sample result could occur under the null

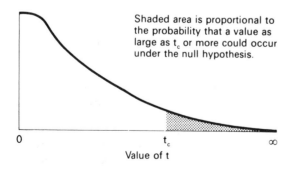

Shaded area is proportional to the probability that a value as large as t_c or more could occur under the null hypothesis.

0   t_c ∞

Value of t

Figure 1.7 Sampling distribution of Students t

hypothesis. It would be possible to reject the null hypothesis as incorrect, knowing that the probability of being wrong in doing so is 'acceptably low', i.e. less than the chosen significance level.

Critical values have been calculated at various significance levels for all the major statistical tests. Appendix C contains critical values for the particular tests described in this book. In general the calculated value of the test statistic must be equal to, or greater than, the critical value in order to permit the null hypothesis to be rejected, but this is not an invariable rule. Details of how to use the tables will be given as each test is discussed.

Degrees of Freedom

It should seem reasonable to suggest that the smaller the samples are, the more difficult it will be to decide whether they accurately reflect the 'real' (population) situation. For smaller samples larger critical values of the test statistic will need to be reached before the null hypothesis can be rejected at any particular significance level.

Degrees of freedom is a number which represents in some way the size of the sample, or samples, involved in the test. In some cases it is simply the sample size, in others it is a value which has to be calculated. Just how this is done, and the logic behind the calculation, will be explained separately for each test in the appropriate sections.

As an example, if the table of critical values of t (Appendix C4) is examined it will be seen that the table has one row of critical values for each number of degrees of freedom. At the 0.1 significance level, the critical value of t is 2.01 with 5 degrees of freedom, 1.81 with 10 degrees of freedom, 1.73 with 20 degrees of freedom, and so on.

Summary

The choice of a particular statistical test depends on the nature of the problem being studied, and on the nature of the data. Tests are available to cope with a variety of situations: comparisons between samples, relationships between variables, trends in time and space. Many of the tests are, however, applicable to only one type of data, whether nominal, ordinal or interval. It is necessary either to choose a test appropriate to the scale of measurement of the data, or convert the data from one scale to another (as discussed in Section 1.2). Many *parametric* tests can only be validly applied to samples taken from populations with a normal distribution, and the results of **all** inferential tests are only reliable if they are applied to unbiased samples.

The choice of an appropriate test determines the *null hypothesis* to be set up. This states that the difference or relationship, or whatever, apparent in the sample data has occurred merely as a result of chance in the sampling process and does not reflect a 'real' difference or relationship in the population. The researcher hopes that his application of the test will enable the null hypothesis to be rejected in favour of the alternative hypothesis.

A *significance level* is decided upon, which is the risk the researcher is prepared to tolerate of wrongly rejecting the null hypothesis. The significance level is a probability, which is unlikely to be greater than 0.1, and may be as low as 0.01 or 0.001. If the researcher subsequently rejects the null hypothesis at the 0.01 level, this means that he is satisfied that the probability that the null hypothesis is in fact correct is no greater than 0.01 (1 chance in 100).

From a knowledge of the *sampling distribution* of the *test statistic* the *critical value* at the specified significance level can be determined. If the significance level is 0.01, for example, the critical value is that value of the test statistic which has a probability of 0.01, of being exceeded by chance under the null hypothesis (i.e. in a situation in which the null hypothesis is in fact true). The critical value also depends on the *degrees of freedom*, which make allowance for the effect of sample size. A certain amount of arithmetic is necessary in arriving at a value of the test statistic, which provides a description of the degree of difference, or whatever, in the sample data. This calculated value of the test statistic can be compared with the critical value. (tables of critical values are given in Appendix C of this book). If the calculated value exceeds the critical value at the 0.01 significance level, there is a probability of no more than 0.01 (1 chance in 100) that such a large value of the test statistic could occur by chance under the null hypothesis. The null hypothesis can be rejected at the 0.01 level, meaning that the chance of being wrong in doing so, is no more than 1 in 100.

Simulating a Statistical Test

As a final demonstration of the way a statistical test works, a simulation of one particular test will be discussed. Simulation is becoming a familiar technique in geography. It is, for example, discussed in both the companion volumes to this book. The type of simulation to be used here is often known as Monte Carlo simulation. It involves the use of random numbers to simulate a large number of possible sample outcomes.

Simulating a statistical test involves the creation of a situation in which the null hypothesis of that test is known to be true. Random numbers are then used to take a sample, or samples, of a specified size from a population whose characteristics satisfy the assumptions of the test. The value of the test statistic can be calculated from this sample data. By repeating the procedure a large number of times, a picture of the sampling distribution of the test statistic under the null hypothesis can be built up. It is also possible to study the effects of taking samples of different sizes from the population, illustrating the influence of degrees of freedom on the sampling distribution.

The simulation of Student's t test (discussed in detail in Section 4.4) provides a fairly simple example of the way this procedure works. The null hypothesis of the t test is that we are dealing with two random samples taken from a common, normally distributed population. Any apparent difference between the two samples is due purely to chance in the sampling process. This null hypothesis situation is created 'artificially' in the simulation by programming a computer to generate a number of pairs of samples taken at random from a normal distribution. For each pair of samples t is calculated in the usual way (see Section 4.4), and a frequency distribution of t values is produced.

Figure 1.8 is an example of the frequency distribution of 1,000 values of t produced by the simulation program, using a sample size of 6 (in a t test two samples of 6 gives 10 degrees of freedom, as explained earlier in this section and in Section 4.4). This process produces a sample

of 1,000 of the possible values of t that could occur under the null hypothesis. From Figure 1.8 it can be seen that there is only one value greater than 4.8. In other words, according to the simulation there is a probability of 0.001 (1/1,000) of getting a value of t greater than 4.8. By experiment the critical value of t at the 0.001 significance level with 10 degrees of freedom has been estimated. This can be compared with the theoretical critical value (Appendix C4), which is 4.59. In the same way, the simulation provides estimates of the critical values of t at other significance levels. Table 1.5 lists the simulated and theoretical critical values side by side for comparison.

Table 1.5 Theoretical and Simulated Critical values of Student's t with 10 degrees of Freedom

Significance level	Theoretical t	Simulated t
0.9	0.13	0.1
0.8	0.26	0.2
0.6	0.54	0.5
0.5	0.70	0.6
0.4	0.88	0.8
0.2	1.37	1.3
0.1	1.81	1.8
0.05	2.23	2.4
0.02	2.76	2.9
0.01	3.17	3.5
0.001	4.59	4.8

The correspondence between the simulated and tabled values of t is good, but not perfect. Hopefully, the reader will be prepared to accept that the simulated sampling distribution of Student's t would get more and more like the tabled one as the number of pairs of samples used in the simulation was increased. The justification for this assertion is that the tabled sampling distribution of Student's t is the distribution of values of t that would be obtained if all possible pairs of random samples of the specified size were drawn from a normally distributed population. All the possible pairs of samples would be an infinite number. As more samples are included in the simulation, this ideal comes closer to being realized.

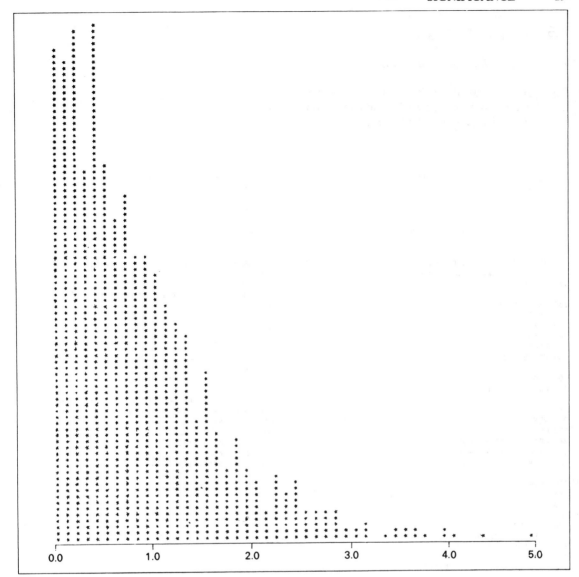

Figure 1.8 Simulated sampling distribution of Student's t with 10 degrees of freedom

The aim of the simulation exercise is to demonstrate how the sampling distribution of a test statistic can be derived experimentally. In fact the critical values given in tables are the result of some fairly complicated mathematics, but the simulation technique can be useful for calculating critical values of newly invented statistics, or of statistics under special conditions. For example it could be used to estimate critical values of a parametric statistic when it is applied to samples taken from a population which is not normally distributed.

1.5 Computer Program

Program 1 — Poisson Probabilities

This program calculates Poisson probabilities
and corresponding expected frequencies from
an observed frequency distribution.

Program Listing

```
10 REM - POISSON PROBABILITIES
20 PRINT"--------------------"
30 PRINT"POISSON PROBABILITIES"
40 PRINT"--------------------"
50 PRINT
60 INPUT"Maximum number of occurrences",N
70 DIM X(N)
80 PRINT
90 PRINT"Now type in frequency of each occurrence:"
100 PRINT"(Mistakes can be corrected later)"
110 FOR I=0 TO N
120 PRINT I;" ";
130 INPUT X(I)
140 NEXT I
150 N1=0
160 N2=N
170 V$=""
180 GOSUB 550
190 S=0
200 T=0
210 FOR I=0 TO N
220 S=S+X(I)
230 T=T+X(I)*I
240 NEXT I
250 PRINT"--------------------"
260 PRINT"POISSON PROBABILITIES"
270 PRINT"--------------------"
280 PRINT
290 PRINT"PROBABILITIES:"
300 PRINT
310 PRINT"      Observed        Poisson"
320 PRINT
330 Z=T/S
340 P=1/EXP(Z)
350 PRINT"0";TAB(6);X(0)/S;TAB(20);P
360 FOR I=1 TO N
370 P=P*Z/I
380 PRINT;I;TAB(6);X(I)/S;TAB(20);P
390 NEXT I
400 PRINT
```

```
410 PRINT"Press SPACE BAR for more..."
420 A$=GET$
430 PRINT
440 PRINT"FREQUENCIES:"
450 PRINT
460 PRINT"      Observed        Poisson"
470 PRINT
480 P=1/EXP(Z)
490 PRINT"0";TAB(6);X(0);TAB(20);P*S
500 FOR I=1 TO N
510 P=P*Z/I
520 PRINT;I;TAB(6);X(I);TAB(20);P*S
530 NEXT I
540 STOP
550 REM - DATA CHECKING ROUTINE
560 PRINT
570 PRINT"Do you want to check/correct ";V$;
580 INPUT" values (Y/N)";Q$
590 IF Q$="N" GOTO 700
600 IF Q$<>"Y" GOTO 570
610 K=0
620 PRINT"Press RETURN if value is correct";
630 PRINT" - otherwise type correct value"
640 FOR I=N1 TO N2
650 K=K+1
660 PRINT K;" ";X(I);
670 INPUT A$
680 IF A$<>"" X(I)=VAL(A$)
690 NEXT I
700 RETURN
```

Example Output

```
---------------------
POISSON PROBABILITIES
---------------------


Maximum number of occurrences?4

Now type in frequency of each occurrence:
(Mistakes can be corrected later)
        0 ?109
        1 ?65
        2 ?22
        3 ?3
        4 ?1

Do you want to check/correct  values (Y/N)?N

---------------------
POISSON PROBABILITIES
---------------------

PROBABILITIES:

        Observed     Poisson

0       0.545        0.543350869
1       0.325        0.33144403
2       0.11         0.101090429
3       1.5E-2       2.0555054E-2
4       5E-3         3.13464573E-3

Press SPACE BAR for more...

FREQUENCIES:

        Observed     Poisson

0       109          108.670174
1       65           66.2888061
2       22           20.2180859
3       3            4.11101079
4       1            0.626929146

STOP at line 540
```

The data used are from the example concerning
deaths by horse-kicks (Table 1.3).

2 Description

Certainly the least contentious function of statistics is to provide concise, and consistent, descriptions of sets of data: concise in that a few numbers can be used to describe the major characteristics of a set of data, and consistent in that people in different parts of the world would calculate these measures in exactly the same way. Some of these measures are in common everyday use, although they may be used rather loosely. For example, people talk about 'average' prices, 'typical' weather, and 'varied' results. These sorts of concept need refining before they can be useful to a statistical geographer.

First of all it might be helpful to consider in a general way what is meant by the 'characteristics of a set of data'. Three types of characteristics will be considered here. These are *central tendency* ('average'), *dispersion* ('spread') and the shape of the frequency distribution. Many of these characteristics can also be measured for spatial data (Section 7.1). For the moment they will only be considered in relation to non-spatial data sets.

2.1 Central Tendency

Measures of *central tendency* are concerned with the 'average' of a set of data. However, 'average' is a very imprecise term. There are numerous ways of calculating an average, each of which has a different interpretation, and not all measures of central tendency are applicable to all types of data.

Mode

The *mode* is the simplest measure of average. The mode, or modal value, is the value which occurs most frequently in the set of data. It is not always very useful as a measure of the average of a set of interval measurements, since any specific value may not occur more than once. Similarly, the occurrence of tied ranks in a set of ordinal data is unlikely to be very common, so the mode will not always be applicable to ordinal data. For nominal data, however, the mode is the most appropriate measure of central tendency. When dealing with nominal data the class which contains the most items (has the highest frequency) is defined as the *modal class*.

Median

The median is another simple measure of 'average'. For a set of data it can be defined as the 'Middle' value of the set when the values are arranged in order. To find the median of the following set of data:

$$4.1, 5.2, 7.3, 2.6, 1.7, 9.4, 8.4$$

they are first put in order:

$$1.7, 2.6, 4.1, 5.2, 7.3, 8.4, 9.4$$

The middle value, 5.2, is the median of this set of data. With an odd number of values in the data set, as in this example, the median is a very simple statistic to work out. For a set of data containing an even number of values the median is simply the value mid-way between the two 'middle' values.

To find the median of the following set of data:

$$7, 2, 7, 4, 8, 5$$

put them in order:

$$2, 4, 5, 7, 7, 8$$

Since there is an even number of values the median is mid-way between the 'middle' two, i.e. 5 and 7. So the median in this case is 6. If the mid-way value is not as obvious as this, it can be easily calculated by adding the middle two values together and dividing by two (in fact by finding the arithmetic mean of the middle two values).

$$(5+7)/2 = 12/2 = 6$$

To find the median of the following set of data (already in order):

$$2.86, 6.97, 7.71, 10.23, 15.95, 17.23$$

The median in this case is the sum of the middle two values divided by two, i.e. $(7.71+10.23)/2 = 17.94/2 = 8.97$.

Arithmetic mean

The *arithmetic mean* (which will from here on be referred to simply as the mean) is a measure of 'average' which can only be applied to interval data. The mean is in fact the measure which is usually implied by the term 'average'. It is calculated by adding together all the values in a data set and dividing by the number of values.

The equation for the mean is:

$$\bar{x} = \frac{\sum x}{n} \tag{2.1}$$

where $\bar{x}$ (read as 'x bar') is the symbol for the mean of variable x. The summation sign $\sum$ (read as 'sigma') means 'the sum of'. Therefore $\sum x$ means 'the sum of all the values of the variable x'. The n in equation 2.1 stands for the number of values of x. For example, the mean of the following set of values:

$$6.2, 9.3, 4.8, 7.2, 5.5$$

is calculated as:

$$(6.2+9.3+4.8+7.2+5.5)/5 = 33.0/5 = 6.6$$

Although the mean is the most commonly used measure of central tendency, it may not be intuitively obvious just what it implies. One possible analogy is to think of it as the 'centre of gravity' of a set of data. This analogy is illustrated in Figure 2.1, applied to the data set used in the above example. Each individual in the data set is depicted as a weight hanging from a weightless ruler. All the weights are equal. The distances between the weights correspond to the intervals between the values in the data set. If such a contraption could be constructed, it would balance at a point equivalent to a value of 6.6 on the ruler.

The problem in building this physical model of the mean in action is clearly to obtain a weightless ruler! However, if the ruler were made as light as possible (say from a piece of aluminium tubing) and the weights as heavy as possible (as heavy as the ruler would bear), then a device like this could actually be used to

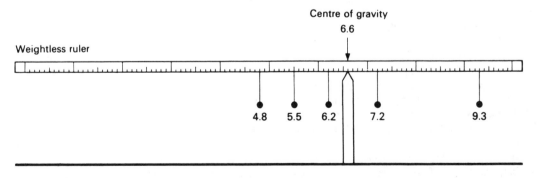

Figure 2.1 The mean as a centre of gravity

measure the approximate mean of a set of values. The heavier the weights, relative to the ruler, the more accurate the device would be.

Another way to think of the mean is as the value which would be obtained by sharing out the values in a data set equally between all the individuals. For example, the mean wage of all British workers is the wage that each one would receive if the total wage bill were shared out equally between all of them.

It is important to realize that mean, median and mode are three quite different measures of central tendency. Not only can they give three different values when they are applied to a common data set, but also their logical bases are quite different. The reader should be quite clear about just what can be inferred from each of these measures.

Exercise 2.1

Calculate the mode, median and mean of each of the following variables from the data matrix (Appendix E):
  (a) Family planning expenditure
  (b) Child centre expenditure
  (c) Working males in social classes I and II.
Consider the implications of each measure and the information it does, or does not, give about the 'average' expenditure. Which seems to be the most informative measure?

2.2 Dispersion

Measures of *dispersion* are concerned with the 'spread' of values in a set of data. In some ways a measure of dispersion is an essential qualification of the 'average' (central tendency) of a set of data. The 'average' may not be a very meaningful measure if it is applied to a set of data whose values are very widely spread. The mean population size of the 50 towns listed in Appendix E is 230,000, but it may be more helpful to know that they vary in fact between 103,261 (Gateshead) and 7,452,346 (London). On its own, the mean in this case is not very informative. Several measures of dispersion are available, and a few of the more common ones will be discussed here.

Range

The range of a set of data is the difference between the smallest and largest values in the data. It is the simplest of all the measures of dispersion. The range of the population size variable in the data matrix is thus $7,452,346 - 103,261 = 7,349,085$. Note that the range is rather misleading in this case. In fact there are only two towns out of the 50 which have populations greater than 1,000,000, the remaining 48 have populations well below this figure. It could be said that the range in this case is 'inflated' in particular by one 'atypical' town.

Quartiles

One way round the problem of 'freak' high or low values which distort the range of a variable is to divide the data set into four parts (*quartiles*), each containing the same number of values. Partition into two parts is achieved by finding the median and using it to classify values as above or below the median. For example, the median of the percentage of working males in social classes I and II (Appendix E) is 7.65. Each town can now be put into one of two classses according to whether is has more than or less than 7.65% of its working males in social classes I and II. Each of these two classes is further subdivided on the basis of the median for the class.

This process is illustrated in Figure 2.2. It provides three 'reference points' in the data set, known as the upper quartile, median and lower quartile. The upper quartile and lower quartile enclose the 'middle half' of the values in the data set. The difference between the upper quartile and the lower quartile is known as the *inter-quartile range*, and it is this measure which can be used instead of the range.

For the social class variable the upper and lower quartiles are 8.9 and 6.9 respectively. The inter-quartile range is therefore $8.9 - 6.9 = 2.0$. The middle half of the values in the distribution fall between 6.9 and 8.9, an inter-quartile range of only 2.0. This can be compared with the range of the same variable, which is 17.0. Note that if the values were evenly spread inter-quartile range would be half the range.

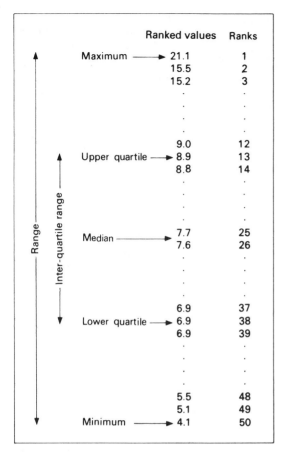

Figure 2.2 Some simple measures of dispersion

Exercise 2.2

Calculate the range, upper and lower quartiles and inter-quartile range of the following variables (Appendix E):
  (a) Population over 65 years (variable C)
  (b) Children in care (Variable E)
  (c) Three car households (variable I)

Mean deviation

One disadvantage of range and inter-quartile range as measures of the spread of a set of data, is that they both take into account only a few values in the data set. They do not consider every single value. In the example illustrated in Figure 2.2, the inter-quartile range is influenced only by the 13th and 38th values in the data set (numbering them from lowest to highest). The 13th and 38th values determine the lower and upper quartiles respectively. The calculation of inter-quartile range provides no information at all about the spread of the other values.

A number of possible measures of dispersion which do take into account every value in the data are based on the deviations of the values from the mean, i.e. the differences between each value and the mean. The following values are measurements of a particular variable for 8 individuals (a to h):

a b c d e f g h

9 2 7 5 4 6 5 2

The mean of these values is 5. Each value can now be expressed as a deviation from the mean by subtracting the mean from it. The value for individual b, for example, becomes $2-5 = -3$. The data now looks like this:

a   b c d e f g h

4 -3 2 0 -1 1 0 -3

or, if the negative signs are ignored (i.e. the *absolute* deviations are considered):

a b c d e f g h

4 3 2 0 1 1 0 3

These values are the original data expressed as absolute deviations from the mean. The mean of these absolute deviations can be calculated as:

$$(4+3+2+0+1+1+0+3)/8 = 1.75$$

This measure is known as the *mean deviation* of a set of data. Its calculation can be summarized in the following equation:

$$\text{mean deviation} = \frac{\sum |x-\bar{x}|}{n} \qquad (2.2)$$

where $|x-\bar{x}|$ is the absolute difference between each value of x and the mean. The summation sign $\sum$ means that the sum of all these absolute deviations is required, and n is the number of values in the data set.

Mean deviation is a fairly simple measure to calculate and to understand, and it does provide a convenient summary of the dispersion of a set of data based on all the values, rather than only a few.

The steps in calculating the mean deviation are illustrated in Figure 2.3 (a and b), so that the geometrical implications of the mean deviation can be appreciated. Mean deviation is not widely used in geography. It is discussed here as an introduction to two more common measures of dispersion, variance and standard deviation.

Variance

In the calculation of mean deviation each of the values in the data set influences the mean deviation in a straight-forward way. An individual which has a value close to the mean contributes little to the mean deviation, one which is further away from the mean contributes more. In fact the contribution made to the mean deviation by any individual in the data set is directly proportional to the difference between the mean and the value for that individual.

An alternative approach is to argue that values further from the mean should have a disproportionately large influence on the measure of dispersion. Consider the three sets of data illustrated in Figure 2.4. Each is represented as a set of deviation from the mean. From looking at them it seems reasonable to suggest that set c has a much greater spread than set a, and that set b is an intermediate situation. However, all three have the same mean deviation. One way of emphasizing the large deviations at the expense of the small ones is to square all the deviations. The effect of this is illustrated in

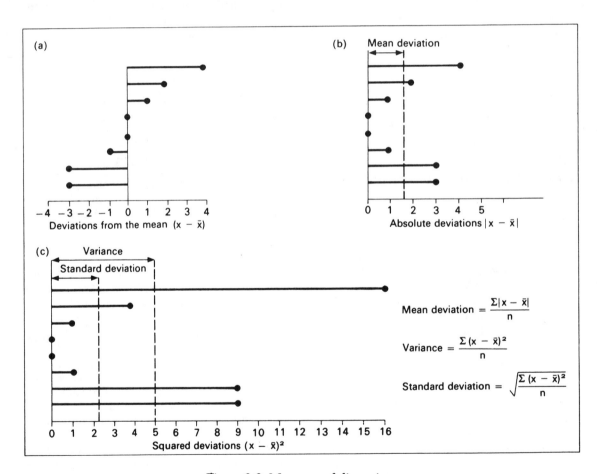

$$\text{Mean deviation} = \frac{\Sigma |x - \bar{x}|}{n}$$

$$\text{Variance} = \frac{\Sigma (x - \bar{x})^2}{n}$$

$$\text{Standard deviation} = \sqrt{\frac{\Sigma (x - \bar{x})^2}{n}}$$

Figure 2.3 Measures of dispersion

Figure 2.3c. Notice that deviations of exactly 1.0 have stayed the same, while the larger deviations have been increased.

By calculating the arithmetic mean of these squared deviations a new measure of dispersion is obtained, known as the *mean square deviation*, or more commonly *variance*. The equation for variance is therefore:

$$\sigma^2 = \frac{\sum (x - \bar{x})^2}{n} \qquad (2.3)$$

where σ^2 (read as 'sigma squared') is the usual symbol for variance, $\sum (x - \bar{x})^2$ is the sum of the squares of all the deviations from the mean, and n is the number of values in the data set.

Equation 2.3 is not particularly quick to use when calculating variance by manual methods, although it is a useful equation for showing what variance implies. It is a helpful *definitional* equation. A better *computational* equation is:

$$\sigma^2 = \frac{\sum x^2}{n} - \bar{x}^2 \qquad (2.4)$$

where $\sum x^2$ means the sum of all the squared values of x, and $\bar{x}^2$ is the square of the mean of x. The calculation of variance is illustrated in Figure 2.3c so that the difference between variance and mean deviation can be appreciated.

Variance is an important measure in statistics, particularly in assessing variation between two or more samples. One very powerful statis-tical technique, known as *analysis of variance*, uses variance to help decide whether a number of samples differ significantly from each other. This technique is discussed in Section 4.7.

Standard deviation

Variance is not often used as a descriptive meas-ure of dispersion. Instead the square root of the variance is taken. This measure is known as the *root mean square deviation*, or simply the *standard deviation*. The definitional equation for the standard deviation is:

$$\sigma = \sqrt{\frac{\sum (x - \bar{x})^2}{n}} \qquad (2.5)$$

where σ (the Greek letter 'sigma') is the symbol for standard deviation. As with variance, there is a simpler computational equation for the standard deviation:

$$\sigma = \sqrt{\frac{\sum x^2}{n} - \bar{x}^2} \qquad (2.6)$$

As an example, Table 2.1 shows the calculation of the standard deviation of a small set of data. The 'mechanics' of standard deviation are also illustrated in Figure 2.3c, for comparison with mean deviation and variance.

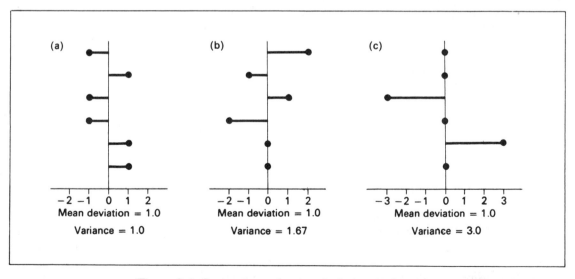

Figure 2.4 Comparison of mean deviation and variance

Table 2.1 Calculation of Standard Deviation

x	x^2
3	9
2	4
1	1
2	4
3	9
4	16
3	9
7	49
6	36
5	25
$\sum x = 36$	$\sum x^2 = 162$

$$\bar{x} = \frac{\sum x}{n} = \frac{36}{10} = 3.6$$

$$\sigma = \sqrt{\frac{\sum x^2}{n} - \bar{x}^2}$$

$$= \sqrt{\frac{162}{10} - 3.6^2}$$

$$= \sqrt{16.2 - 12.96}$$

$$= \sqrt{3.24} = 1.8$$

One simple advantage of standard deviation over variance is that the standard deviation of a set of values is almost always considerably less than the variance. The variance of a set of data can easily be a very large number, larger even than any of the individual values in the data set. This is a direct consequence of all the squarings involved in the calculation of variance. By taking the square root of variance to produce standard deviation a more 'manageable' number results.

The standard deviation has another very useful property. For any data set the following relationships hold:

(1) at least 56% of the individuals have values which are within 1.5 standard deviations either side of the mean (i.e. values between $\bar{x} - 1.5\sigma$ and $\bar{x} + 1.5\sigma$)

(2) at least 75% of individuals have values which are within two standard deviations either side of the mean (i.e. values between $\bar{x} - 2\sigma$ and $\bar{x} + 2\sigma$)

(3) at least 89% of individuals have values which are within three standard deviations either side of the mean (i.e. values between $\bar{x} - 3\sigma$ and $\bar{x} + 3\sigma$)

(4) at least 94% of individuals have values which are within four standard deviations either side of the mean (i.e. values between $\bar{x} - 4\sigma$ and $\bar{x} + 4\sigma$.

These relationships are always true, regardless of the shape of the frequency distribution of the data. They are derived from Tchebysheff's theorem.

Standard deviation is the measure of dispersion most commonly applied to geographical data. It also forms the basis of a number of parametric statistical methods concerned with making estimates from samples. Some of these methods will be discussed in the next chapter.

Exercise 2.3

Calculate the standard deviation of the percentage of working males in social classes I and II for the 50 towns in Appendix E.

2.3 Skewness and Kurtosis

Standard deviation and other measures of dispersion are concerned with the spread of values in a frequency distribution. In a sense they measure the 'width' of the distribution. However, measures of dispersion do not provide any information about other characteristics of the shape of a frequency distribution. Figure 2.5 shows six frequency distributions in the form of histograms. These frequency distributions represent sets of data containing the same number of values. The horizontal scales of the histograms have been standardized by expressing

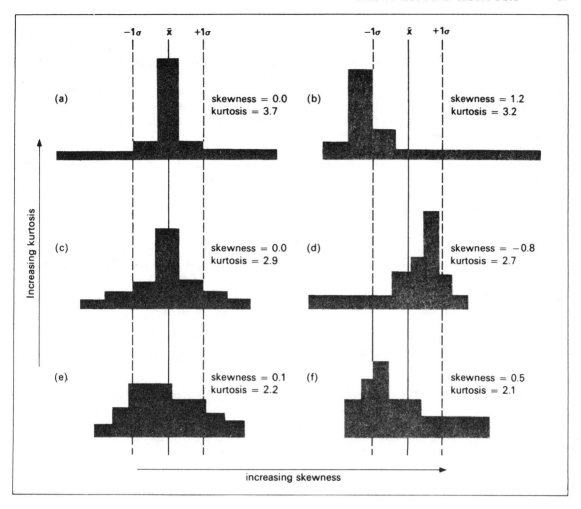

Figure 2.5 Skewness and kurtosis

them in units of standard deviation on either side of the mean, and the areas of the histograms have been made the same. Direct comparisons can, therefore, be made between them.

It is apparent that the six frequency distributions are very different. Those on the right hand side of figure 2.5 (b, d, and f) are, however, similar in one respect: they are markedly asymmetrical. The 'peak', or mode, of each of these three distributions is displaced to one side of the mean. They are *skewed* distributions. The three distributions in the left hand half of Figure 2.5 (a, c, and e) have a low degree of *skewness*. In each case the mode is very close to, or even

coincides with, the mean. They are all symmetrical distributions.

The two distributions on each row can be considered similar in the sense that they have a similar degree of 'peakiness' or *kurtosis*. The two distributions on the top row of figure 2.5 (a and b) are very 'peaky'; they have a high degree of kurtosis. Such distributions are sometimes referred to as *leptokurtic* distributions. The distributions in the second row (c. and d) are *mesokurtic*; they have a moderate degree of 'peakiness'. The bottom row shows two *platykurtic*, or relatively 'flat', distributions (e and f).

Skewness

Skewness, then, measures the extent to which the bulk of the values in a distribution are concentrated to one side or the other of the mean. If the bulk of the values are less than the mean, the distribution is said to be *positively skewed* (e.g. Figure 2.5b). If there are more values greater than the mean, the distribution is *negatively skewed* (e.g. Figure 2.5d). A perfectly symmetrical distribution, such as figure 2.5c has no skewness.

There are various statistical measures of skewness. The most common, properly known as *momental* skewness, is calculated using the following equation:

$$\text{Skewness} = \frac{\sum (x - \bar{x})^3}{n\sigma^3} \qquad (2.7)$$

where $(x - \bar{x})^3$ denotes the cube of the deviations of the values from their mean, σ is the standard deviation, and n is the number of values. The application of Equation 2.7 to a simple data set is shown in Table 2.2. The value of skewness for a symmetrical distribution is zero. Logically enough, negative values of the index indicate negative skewness, and positive values indicate positive skewness.

Skewness is an important concept in geographical statistics because very many of the variables measured in geographical studies have highly skewed distributions. In other words, their frequency distributions look like those shown in Figure 2.5b, d, or f. This fact has two important consequences. First it casts doubt upon the advisability of applying parametric statistical tests to these data. A high degree of skewness is one sign that sample data are not normally distributed. They are therefore unlikely to come from a population which is normally distributed. Remember that a normal distribution (Section 1.3) is symmetrical and therefore has a skewness of zero. The assumptions of parametric tests were discussed in section 1.4.

Secondly, other descriptive measures, particularly the mean, may be misleading if used in

Table 2.2 Calculation of Skewness and Kurtosis

x	$x - \bar{x}$	$(x - \bar{x})^2$	$(x - \bar{x})^3$	$(x - \bar{x})^4$
1	−2.6	6.76	−17.576	45.6976
2	−1.6	2.56	−4.096	6.5536
2	−1.6	2.56	−4.096	6.5536
3	−0.6	0.36	−0.216	0.1296
3	−0.6	0.36	−0.216	0.1296
3	−0.6	0.36	−0.216	0.1296
4	0.4	0.16	0.064	0.0256
5	1.4	1.96	2.744	3.8416
6	2.4	5.76	13.824	33.1776
7	3.4	11.56	39.304	133.6336
36		32.40	29.520	229.8720

$$\bar{x} = \frac{\sum x}{n} = \frac{36}{10} = 3.6 \qquad \sigma = \sqrt{\frac{\sum (x - \bar{x})^2}{n}} = \sqrt{\frac{32.4}{10}} = \sqrt{3.24} = 1.8$$

$$\text{skewness} = \frac{\sum (x - \bar{x})^3}{n\sigma^3} = \frac{29.520}{10 \times 1.8^3} = \frac{29.520}{10 \times 5.832} = \frac{29.520}{58.32} = 0.506$$

$$\text{kurtosis} = \frac{\sum (x - \bar{x})^4}{n\sigma^4} = \frac{229.8720}{10 \times 1.8^4} = \frac{229.8720}{10 \times 10.4976} = \frac{229.8720}{104.976} = 2.190$$

isolation. This can best be demonstrated by example. The mean Local Authority expenditure on family planning in 1971 (Appendix E, variable G) was £7.47 per 1000 population. Taking this figure on its own, it might be assumed that about half the L.A.s will have spent more and half less than £7.47. In fact, however, only a quarter of the L.A.s spent more than this figure. Well over half spent nothing at all on family planning in 1971. In this rather extreme case the mean on its own is not a very informative measure. Drawing a histogram of the data would soon bring to light the fact that this is a highly skewed distribution. Equation 2.7 can be used to quantify the skewness and thus enable objective comparisons to be made with other distributions.

Kurtosis

Kurtosis measures the extent to which values are concentrated in one part of a frequency distribution. If one class, or group of adjacent classes, in a frequency distribution contains a large proportion of all the values in the distribution, then the distribution has a high degree of kurtosis. It is very 'peaky' or *leptokurtic*. In a distribution with a low degree of kurtosis (a 'flat', or *platykurtic* distribution) each class contains a similar proportion of all the values.

The usual measure of kurtosis is calculated with the following equation:

$$\text{kurtosis} = \frac{\sum (x - \bar{x})^4}{n\sigma^4} \qquad (2.8)$$

where $(x - \bar{x})^4$ denotes the fourth power of the deviations of the values from their mean, σ is the standard deviation, and n is the number of values. The application of equation 2.8 to a simple data set is shown in Table 2.2, alongside the calculation of skewness.

The particular measure of kurtosis given by Equation 2.8 is devised in such a way that a normal distribution (Section 1.3) has a kurtosis of 3.0. A very 'peaky' (leptokurtic) distribution has a kurtosis greater than 3.0. A very 'flat' (platykurtic) distribution has a kurtosis less than 3.0.

Kurtosis is not a widely used measure in geographical studies, certainly not as widely used as it should be. Like skewness, it gives valuable information about the distribution of a set of data values, in addition to that provided by the mean and standard deviation. Several of the variables in Appendix E are highly leptokurtic, for example, per capita expenditure on child care (kurtosis = 8.98), and percentage of working males in classes I and II (kurtosis = 7.76).

It is also worth noting that many geographical variables are both highly skewed **and** highly leptokurtic. When applied to these sorts of variable, mean and standard deviation may give quite misleading impressions. Furthermore, such data, if they relate to samples, are very unlikely to come from normally distributed populations, making them unsuitable for parametric tests. This point was discussed in Section 1.4, and will be made again several times in later chapters. Remember that a normal distribution has a skewness of zero and a kurtosis (measured using Equation 2.8) of 3.0.

Exercise 2.4

(a) Calculate the skewness and kurtosis of the following set of values:

$$2, 2, 1, 3, 4, 7, 3, 2, 4, 2$$

(b) Calculate the skewness and kurtosis of variable E (children in care) from Appendix E.

2.4 Computer Program

Program 2 — Descriptive Statistics

The program calculates most of the descriptive statistics discussed in this chapter, as well as the best estimate of the standard deviation, which is explained in Section 3.2.

Program Listing

```
10 REM - DESCRIPTIVE STATISTICS
20 PRINT"----------------------"
30 PRINT"DESCRIPTIVE STATISTICS"
40 PRINT"----------------------"
50 PRINT
60 INPUT"Number of values",N
70 DIM X(N),N(N),R(N)
80 PRINT
90 PRINT"Now type in data values:"
100 PRINT"(mistakes can be corrected later)"
110 FOR I=1 TO N
120 PRINTI;" ";
130 INPUT X(I)
140 N(I)=I
150 NEXT I
160 N1=1
170 N2=N
180 V$="data"
190 GOSUB 900
200 GOSUB 730
210 XM=0
220 FOR I=1 TO N
230 XM=XM+X(I)
240 NEXT I
250 XM=XM/N
260 X2=0
270 X3=0
280 X4=0
290 FOR I=1 TO N
300 X2=X2+X(I)^2
310 X3=X3+(X(I)-XM)^3
320 X4=X4+(X(I)-XM)^4
330 NEXT I
340 V=X2/N-XM^2
350 SD=SQR(V)
360 SB=SD*SQR(N/(N-1))
370 SK=X3/(N*SD^3)
380 KT=X4/(N*V^2)
390 J=(N-1)/2
400 K=J/2+1
410 J=J+1
420 L=J-INT(J)
430 J=INT(J)
440 MD=X(J)+(X(J+1)-X(J))*L
450 L1=INT(K)
460 L2=L1+1
470 LQ=X(L1)+(X(L2)-X(L1))*(K-INT(K))
480 K=N-K+1
490 L1=INT(K)
500 L2=L1+1
510 UQ=X(L1)+(X(L2)-X(L1))*(K-INT(K))
520 PRINT
530 PRINT"----------------------"
```

```
540 PRINT"DESCRIPTIVE STATISTICS"
550 PRINT"----------------------"
560 PRINT
570 PRINT"Mean = ";XM
580 PRINT"Standard error of mean = ";SB/SQR(N)
590 PRINT"Variance = ";V
600 PRINT"Standard deviation = ";SD
610 PRINT"Best estimate of S.D. = ";SB
620 PRINT"Standard error of S.D. = ";SB/SQR(2*N)
630 PRINT"Skewness = ";SK
640 PRINT"Kurtosis = ";KT
650 PRINT"Minimum = ";X(1)
660 PRINT"Maximum = ";X(N)
670 PRINT"Range = ";X(N)-X(1)
680 PRINT"Median = ";MD
690 PRINT"Lower quartile = ";LQ
700 PRINT"Upper quartile = ";UQ
710 PRINT"Inter-quartile range = ";UQ-LQ
720 STOP
730 REM - BUBBLE SORT RANKING ROUTINE
740 Z=0
750 FOR I=1 TO N-1
760 J=I+1
770 IF X(I)<X(J) GOTO 870
780 IF X(I)>X(J) GOTO 800
790 GOTO 870
800 D=X(J)
810 DN=N(J)
820 N(J)=N(I)
830 X(J)=X(I)
840 N(I)=DN
850 X(I)=D
860 Z=1
870 NEXT I
880 IF Z=1 GOTO 740
890 RETURN
900 REM - DATA CHECKING ROUTINE
910 PRINT
920 PRINT"Do you want to check/correct ";V$;
930 INPUT" values (Y/N)";Q$
940 IF Q$="N" GOTO 1050
950 IF Q$<>"Y" GOTO 920
960 K=0
970 PRINT"Press RETURN if value is correct";
980 PRINT" - otherwise type correct value"
990 FOR I=N1 TO N2
1000 K=K+1
1010 PRINT K;" ";X(I);
1020 INPUT A$
1030 IF A$<>"" X(I)=VAL(A$)
1040 NEXT I
1050 RETURN
```

Example Output

```
----------------------
DESCRIPTIVE STATISTICS
----------------------

Number of values?10

Now type in data values:
(mistakes can be corrected later)
        1 ?2
        2 ?2
        3 ?1
        4 ?3
        5 ?4
        6 ?7
        7 ?3
        8 ?2
        9 ?4
       10 ?2

Do you want to check/correct data values (Y/N)?N

----------------------
DESCRIPTIVE STATISTICS
----------------------

Mean = 3
Standard error of mean = 0.483735465
Variance = 2.6
Standard deviation = 1.61245155
Best estimate of S.D.  = 1.69967317
Standard error of S.D. = 0.342052628
Skewness = 1.28805301
Kurtosis = 4.11242603
Minimum = 1
Maximum = 7
Range = 6
Median = 2.5
Lower quartile = 2
Upper quartile = 3.75
Inter-quartile range = 1.75

STOP at line 720
```

The data used here are from Exercise 2.4a.

3 Samples and Sampling

From time to time there is great interest in the United Kingdom in the result of a coming general election. In the preceding weeks the results of various opinion polls are published, each of which aims to give an accurate prediction of the way in which the electorate as a whole will vote. These polls are based on interviews with only one or two thousand voters out of a total of over 40 million. The choice of sample voters must obviously be made with care to ensure that the sample is truly representative of electoral opinion in the country at large. It would be unrealistic to expect a reasonable estimate of the voting intentions of the British electorate from a sample of interviews held only in the Highlands of Scotland, a Yorkshire mining area or an outer suburb of London.

In this case the reliability of the sample survey is revealed soon after polling day. In other cases, including many geographical ones, the total situation may be so vast that it can never be studied as a whole. A geomorphologist interested in coastal processes may want to investigate differences in the size and shape of sand grains along a beach. The number of sand grains on a beach, though not infinite, is uncountably large. The geomorphologist will have to be content with studying samples of sand. A number of questions need to be answered: how should these samples be chosen, how large should they be, how much reliance can be placed on the sample measurements as estimates of the characteristics of the millions of other sand grains on the beach?

These are questions that the statistician can help to answer. However, before discussing sampling in detail it may be helpful to introduce the topic by means of a practical exercise.

Exercise 3.1

Figure 3.1 is a map of part of a town. The map area includes 200 households distributed in several different districts each of fairly homogeneous socio-economic character. You are a representative of Spoton Surveys and you only have time to visit 50 houses in the area to find the answers to four questions:

(1) Is the head of the household a New Commonwealth immigrant (e.g. West Indian, Pakistani)?

(2) Does the household have television?

(3) Does the household have a car?

(4) Will the head of the household vote for the National Democratic party at the next election?

Various methods of choosing the sample are possible:

(1) with a pin (eyes closed),

(2) by eye,

(3) take every fourth house on the list (Table 3.1),

(4) go down each street picking every fourth house you come to,

(5) place a grid over the map and take an appropriate number of houses from each grid square,

(6) put all 200 addresses in a hat and pick out 50,

(7) any other method you can think of.

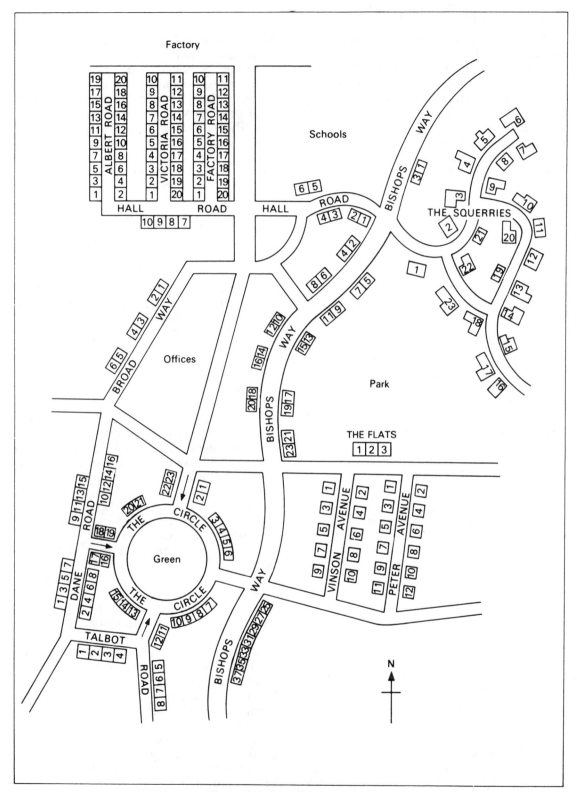

Figure 3.1 Map for exercise 3.1

Choose the 50 households you are to visit and, with the aid of Table 3.1, find the answers to the four questions for each of the 50. Table 3.1 represents the actual situation and contains the answers for each of the 200 households. Do not consult this table until you have decided on your sample.

From your sample results calculate the percentage of households with New Commonwealth immigrants as heads, the percentage with televisions, the percentage with cars and the percentage who will vote National Democratic. Compare your sample percentages with the total percentages (obtained from Table 3.1) which are:

(1) New Commonwealth immigrants
$$15/200 = 7.5\%$$
(2) Television owners
$$160/200 = 80\%$$
(3) Car owners
$$100/200 = 50\%$$
(4) National Democratic voters
$$70/200 = 35\%$$

If there are major discrepancies between your sample results and the 'real' results, can you account for them? If the exercise is being carried out by a whole class, which method of sampling produces the most accurate answers?

3.1 Sampling Methods

Not all methods of sampling are equally suitable for choosing samples which are representative of the population. If the grid square method of selection is used in the previous exercise it is likely to produce a sample which contains too few houses from areas of high housing density and too many houses from areas where housing density is low. This might mean choosing say 10 houses from the Albert Road/Victoria Road/Factory Road area and 10 houses from The Squerries, since they both occupy approximately the same area of land. As there are 60 houses in Albert Road, Victoria Road and Factory Road, the representation rate in this area is 10/60, or approximately 17%. For The Squerries, however, the representation rate is 10/23, or approximately 43%. Compared with the intended sampling rate of 25% (50/200), the Albert Road/Victoria Road/Factory Road area is *underrepresented*, whereas The Squerries

is *overrepresented*. This would have serious repercussions on the results of the survey, since The Squerries has, for example, very high rates of car and television ownership.

The choice of sampling method should be made so as to avoid as far as possible this sort of bias. The remainder of this section will be concerned with discussing the advantages and disadvantages of different sampling methods. A very large number of sampling methods is available, but two elementary distinctions can be made: first between systematic and random methods and secondly between spatial and non-spatial methods. The choice of sampling method depends both on the nature of the situation and on the purpose for which the sample is required. It has already been suggested (in Section 1.4) that random samples are a prerequisite of many inferential statistical techniques. However, for some purely descriptive purposes samples which are not random may be more useful.

Random Sampling

A *random* sampling method is one in which the choice of individuals for inclusion in the sample is left entirely to chance. Picking names from a hat or choosing spots on a map where raindrops pitch are therefore random sampling methods. The choice of any particular individuals is in no way influenced by the researcher at any stage in the sampling process. Once one name or one spot on a map has been picked there is no way of knowing which name or spot will be picked next.

A random sampling method should satisfy two important criteria:

(1) every individual must have an equal chance of inclusion in the sample throughout the sampling procedure,

(2) the selection of any particular individual should not affect the chance of selection of any other individual.

To put these criteria in more formal probability terms: the probabilities of inclusion in the sample must be **equal** and **independent** of each other. So, if the aim is to pick a random sample of 50 households from a population of 200, every household should have the same 50/200 or 0.25 probability of selection.

Table 3.1 Household Data for Exercise 3.1

R	S	I	T	C	N	R	S	I	T	C	N
Albert Road						*Factory Road* (contd.)					
1	1	0	1	0	0	45	5	0	0	0	0
2	3	0	1	0	0	46	6	0	1	0	0
3	5	0	0	0	0	47	7	0	1	0	0
4	7	0	0	1	0	48	8	0	0	1	0
5	9	0	1	0	0	49	9	0	1	0	0
6	11	0	1	0	0	50	10	0	1	0	0
7	13	0	1	0	0	51	11	0	1	1	0
8	15	0	1	0	0	52	12	0	0	0	0
9	17	0	0	1	0	53	13	0	1	0	0
10	19	0	1	0	0	54	14	0	1	0	0
11	2	0	1	0	0	55	15	0	0	0	1
12	4	0	1	1	0	56	16	0	1	0	1
13	6	0	0	1	0	57	17	0	1	0	1
14	8	0	1	0	0	58	18	0	1	0	0
15	10	0	1	0	0	59	19	0	0	0	1
16	12	0	0	0	0	60	20	0	0	0	1
17	14	0	1	0	0	*Hall Road*					
18	16	0	1	0	0	61	1	0	1	0	0
19	18	0	0	0	0	62	2	0	1	0	0
20	20	0	0	0	0	63	3	0	1	1	0
Victoria Road						64	4	0	0	0	0
21	1	0	1	0	0	65	5	0	1	0	0
22	2	0	1	1	0	66	6	0	1	0	1
23	3	0	1	0	0	67	7	1	0	0	0
24	4	0	0	0	0	68	8	1	1	1	0
25	5	0	1	0	0	69	9	0	1	0	1
26	6	0	1	0	0	70	10	0	1	1	1
27	7	0	1	0	0	*Broad Way*					
28	8	0	0	0	0	71	1	0	1	1	0
29	9	0	0	0	0	72	2	0	1	1	1
30	10	0	1	0	0	73	3	0	1	0	1
31	11	0	1	0	0	74	4	0	1	0	0
32	12	0	1	1	0	75	5	0	1	0	1
33	13	0	1	0	0	76	6	0	1	1	1
34	14	0	0	0	0	*Dane Road*					
35	15	0	1	0	0	77	1	0	1	0	1
36	16	0	1	0	0	78	3	0	1	1	1
37	17	0	1	0	0	79	5	0	1	1	1
38	18	0	1	1	0	80	7	0	1	1	0
39	19	0	0	0	0	81	9	0	1	0	0
40	20	0	1	0	0	82	11	0	1	0	1
Factory Road						83	13	0	1	1	1
41	1	0	0	1	0	84	15	0	1	0	1
42	2	0	0	0	0	85	2	0	1	0	1
43	3	0	1	0	0	86	4	0	1	0	0
44	4	0	1	0	0	87	6	0	0	1	0

Table 3.1 continued

R	S	I	T	C	N	R	S	I	T	C	N
Dane Road (contd.)						*Bishops Way* (contd.)					
88	8	0	1	1	1	130	7	0	0	0	1
89	10	0	1	1	0	131	9	0	1	1	1
90	12	0	1	1	1	132	11	0	0	1	1
91	14	0	1	0	0	133	13	0	1	1	1
92	16	0	1	0	1	134	15	0	1	0	1
Talbot Road						135	17	1	0	1	0
93	1	0	1	1	1	136	19	1	1	0	0
94	2	0	1	0	1	137	21	0	0	1	1
95	3	0	1	1	1	138	23	1	0	0	0
96	4	0	1	1	0	139	25	0	1	0	1
97	5	0	1	0	1	140	27	1	1	0	0
98	6	0	1	1	1	141	29	1	0	0	0
99	7	0	1	0	0	142	31	1	1	1	0
100	8	0	1	0	1	143	33	0	1	1	1
The Circle						144	35	1	1	0	0
101	1	0	1	1	0	145	37	1	1	0	0
102	2	0	1	0	1	146	2	1	1	0	0
103	3	0	1	1	1	147	4	1	0	1	0
104	4	0	1	1	1	148	6	0	0	1	1
105	5	0	1	1	0	149	8	0	1	1	1
106	6	0	1	1	1	150	10	0	1	0	1
107	7	0	0	0	0	151	12	0	1	0	1
108	8	0	1	1	1	152	14	1	1	0	1
109	9	0	1	0	0	153	16	1	0	0	0
110	10	0	1	1	1	154	18	0	1	0	1
111	11	0	1	1	0	155	20	0	1	0	0
112	12	0	1	1	1	*The Squerries*					
113	13	0	1	1	0	156	1	0	1	1	0
114	14	0	0	0	0	157	2	0	1	1	0
115	15	0	1	1	1	158	3	0	1	1	0
116	16	0	1	1	0	159	4	0	1	1	0
117	17	0	1	0	1	160	5	0	1	1	0
118	18	0	1	1	0	161	6	0	1	1	0
119	19	0	1	1	0	162	7	0	1	1	0
120	20	0	1	1	0	163	8	0	1	1	0
121	21	0	1	1	0	164	9	0	1	1	0
122	22	0	0	1	0	165	10	0	1	1	0
123	23	0	1	1	0	166	11	0	0	1	1
The Flats						167	12	0	1	1	0
124	1	0	1	1	1	168	13	0	1	1	0
125	2	0	1	1	1	169	14	0	1	1	0
126	3	0	1	0	1	170	15	0	1	1	0
Bishops Way						171	16	0	1	1	0
127	1	0	1	1	1	172	17	0	1	1	0
128	3	1	1	1	0	173	18	0	1	1	0
129	5	0	1	0	1	174	19	0	1	1	0

Table 3.1 continued

R	S	I	T	C	N	R	S	I	T	C	N
The Squerries (contd.)						Vinson Avenue (contd.)					
175	20	0	1	1	0	188	10	0	1	1	1
176	21	0	1	1	0	Peter Avenue					
177	22	0	1	1	0	189	1	0	1	1	0
178	23	0	1	1	0	190	3	0	0	1	1
Vinson Avenue						191	5	0	1	1	0
179	1	0	1	1	1	192	7	0	1	0	1
180	3	0	1	1	1	193	9	0	1	1	1
181	5	0	1	0	0	194	11	0	1	1	1
182	7	0	0	1	1	195	2	0	1	0	0
183	9	0	1	1	0	196	4	0	0	1	1
184	2	0	1	1	0	197	6	0	1	1	1
185	4	0	1	1	1	198	8	0	1	1	0
186	6	0	1	1	0	199	10	0	1	1	0
187	8	0	0	0	1	200	12	0	1	0	1

R = reference number (for sampling with random numbers)
S = street number of house
I = immigrant
T = television
C = car
N = N.D.P. supporter   1 = yes   0 = no

Exercise 3.2

Which of the sampling methods suggested in Exercise 3.1 satisfy the criteria of random sampling?

Satisfying the two criteria is not quite as straightforward as it seems. Suppose the households for Exercise 3.1 are selected by picking 50 addresses from a hat one at a time. At the first 'dip' all the households have an equal chance of selection, in fact a probability of 1/200 or 0.005. If the first address slip is taken out and thrown away, at the next 'dip' the remaining 199 households have a 1/199 chance of selection. Clearly, the household chosen at the first dip now has no chance at all (a probability of 0.0) of selection. This process would continue throughout the sampling so that any household not selected on any occasion would have a slightly increased probability of selection the next time around. By the time the 50th household came to be selected there would be only 151 addresses left in the hat, each with a 1/151 chance of selection.

This alteration of the probabilities as the sampling proceeds clearly violates the first criterion of a random sample. It comes about as a result of sampling *without replacement*. If the individuals are replaced (e.g. the addresses put back into the hat) as soon as each selection has been made, the households will have the same 1/200 probability of inclusion at every stage in the sampling. After all 50 selections have been made each household will have had the same 50 × 1/200, or 0.25, probability of inclusion in the sample.

There is one inescapable consequence of selecting a random sample with replacement which may seem rather strange at first. It is perfectly possible for any individual to be picked for the sample more than once, and even every time. Strictly speaking, if an individual is picked twice (or more) then the data for that individual should be included twice (or however many times) in any subsequent analysis. In practice this is seldom done, and the bias introduced into the study by not doing so is very unlikely to be of any importance.

Exercise 3.3

What is the probability of choosing the same household on every occasion in the random selection of 50 households out of a population of 200?

Random Numbers

Although picking names or numbers from a hat is a perfectly satisfactory way of choosing a random sample, it can be rather time-consuming when large samples are required or large populations are being dealt with. Provided the individuals in the population have been numbered consecutively, a convenient alternative is to use tables of random numbers. Appendix D1 is a table of 1000 pairs of random numbers between 00 and 99. These numbers have been picked at random, in this case using a computer program. They have been selected in such a way that each pair of digits has an equal probability of occurrence. In other words we should expect each of the numbers 00 to 99 to occur approximately the same number of times in the table as a whole. The reader may like to check for himself that this in fact the case. Furthermore, the numbers occur in random order, whether they are read down a column or across a row.

The digits can be used singly to pick a sample from a population of 10 items (numbered 0 to 9), in pairs to pick a sample from a population of 100 (numbered 0 to 99), in triplets to pick a sample from a population of 1000, and so on. Often, however, the geographer wants to pick a sample from a population which does not contain such a convenient number of individuals. It may not be immediately obvious how a table of random numbers can be used to pick a sample from a population of, say, 1473 people named on an electoral register. In this case it would be necessary to use the random digits in groups of four, rejecting any resulting numbers that were larger than 1473 or less than 1.

Note that in general the randomness of random numbers is not destroyed by adding or subtracting a constant, or multiplying or dividing by a constant. More complicated transformations, such as taking squares or square roots, should not be used.

Systematic Sampling

A *systematic* sample is one which is selected in some regular way. In Exercise 3.1, taking every fourth address from the list would have produced a systematic sample. Intuitively, this would seem to produce an even, and therefore 'fair', coverage of the population in the sense that sample members are selected evenly from throughout the list, avoiding the 'bunching' that can often occur with random sampling. Certainly in a systematic sample no individual can be picked more than once.

It may not be immediately obvious that this 'evenness' does not necessarily produce a sample which is unbiased and representative of the population from which it is taken. However, it is not difficult to show that systematic sampling does not give every individual in a population the same chance of selection.

Suppose a 10% systematic sample is being taken from a list of 100 names. If every tenth name is selected starting from number 1, then the 1st, 11th, 21st, 31st etc. name has a probability of 1.0 of selection. For all the other names on the list the probability of selection is 0.0. In other words, once the first selection has been made 90% of the names on the list have lost all chance of representation in the sample.

Systematic sampling has, however, certain practical advantages over random sampling. In particular, it is usually easier and quicker to obtain a systematic sample than a random one. In many geographical situations this may be a major consideration. However, if the decision is taken to use a systematic sample it may not be possible to make reliable inferences about the population on the basis of the sample.

The important point is to ensure that the sampling method does not introduce a consistent bias into the sample. In spatial sampling, for example, there is the danger that a systematic method will pick out regularities in the base. In areas of regular housing layouts, such as Victorian terraces, picking every, say, 20th house might result in selecting only corner houses. Since these are generally more expensive than the other houses in the terrace they might be expected to contain atypical households.

Spatial Sampling

Many geographical studies are concerned with the spatial distribution of a variable, such as altitude or rainfall, which varies continuously from place to place. For a large area it would be impossible to measure the value of this variable at every point. The aim of sampling in this situation is first of all to provide as economically as possible and as accurately as possible a description of the spatial distribution of that variable. The most likely follow-up to this procedure would be the production of a map.

A number of problems need to be considered. What method of sampling is most effective? How large does the sample need to be? Some variables can be measured effectively at points, for example soil acidity (pH) or moisture content. Other variables must be measured over an area, for example vegetation cover, population density or various socioeconomic indices such as literacy, mortality and morbidity (disease rates). For these variables it is also necessary to decide on the most effective size and shape of the sampling areas, or *quadrats*.

In spatial sampling a major distinction can be made between random and systematic methods. The difference from non-spatial sampling is that the aim is to select points or quadrats from a continuous surface rather than items from a list. The simplest method of spatial random sampling consists of using a pair of random numbers to define a pair of co-ordinates (easting and northing) which can be used to locate on a map the points or quadrats that are to make up the sample. Figure 3.2a shows a random sample of 20 points selected in this way from a study area. The co-ordinates of each point were fixed by a pair of random numbers taken in succession from a table of random numbers. Points which fell outside the boundary of the study area were rejected.

Figure 3.2b shows a systematic sample of 20 points selected from the same study area by superimposing a grid of suitable dimensions over the area and letting each grid intersection

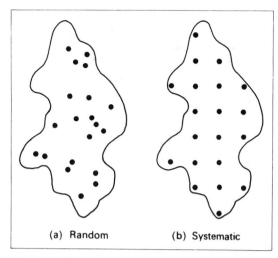

(a) Random   (b) Systematic

Figure 3.2 Sampling methods

define a sample point. It is in fact a very difficult task to design a grid which will give a predetermined number of sample points in an irregular area. In reality a researcher would probably have to be content with a sample containing only approximately the desired number of points.

There are many other ways of obtaining spatial samples. Rather than consider them in detail at this stage, the reader can investigate some of the problems inherent in spatial sampling by trying the exercise which follows.

Exercise 3.4

Figure 3.3 shows an area of coastline backed by salt marsh. You are a biogeographer carrying out research into the effects of salinity on salt marsh vegetation. Vegetation has been accurately mapped for the whole of the area, perhaps using information from aerial photographs as well as field survey. Your task is to produce a map of salinity for comparison with the vegetation map. Time and money are both limited so that, in the first instance at least, you can only afford to make 50 measurements of salinity.

Decide on a method of sampling, then pick 50 points inside the study area at which to take measurements. For the purpose of this exercise measurements are made by consulting Figure 3.4, which shows the salinity levels for the whole

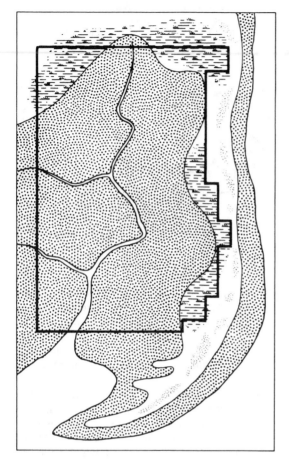

Figure 3.3 Salt marsh, showing boundary of study area

average rainfall for a drainage basin, then random sampling is necessary in order to satisfy the assumptions of the statistical techniques employed in calculating the reliability of these estimates. Such techniques are discussed in Section 3.2.

More often the geographer is concerned with the change in a continuous variable over an area, usually in response to, or at any rate in conjunction with, changes in a number of other variables. In such a situation the major concern of the geographer is that he should not miss any major parts of the overall 'pattern' of the distribution of the variable or variables under study. This may dictate relying more on common sense than on any theoretical considerations. In the case of Exercise 3.4, for example, a systematic sample is probably the more efficient from the point of view of obtaining an overall picture of salinity levels. There may be parts of the study area where this regular grid of sample points is too coarse to enable an isoline map to be drawn with any degree of confidence. In these sub-areas more detailed sampling with a finer grid may be necessary to enable the researcher to 'see' the pattern clearly.

One inherent danger of systematic sampling is that it is possible to miss entirely elements of a complex regular pattern, if these elements happen to occur 'in phase' with the sampling grid. Figure 3.5 illustrates such a situation. This is an isoline map of soil acidity (pH) in an area where environmental conditions have produced a remarkably regular pattern, with patches of high acidity alternating with less acid patches in a systematic way over the whole of the area. If a sampling exercise were to be carried out using the points marked on Figure 3.5, the results would suggest remarkably little variation in acidity over the area.

In the case of soil acidity such 'patterning', depending on the scale at which it occurred, would normally be visible in vegetation differences. This need not always be the case, however, and certainly there are other continuously distributed variables which do not produce visible signs of their variation. Methods for detecting patterning in continuous spatial distributions have been developed, particularly by plant ecologists. For a review of some of these methods the interested reader should consult Kershaw (1964).

area. If you use a tracing of the map in your sampling procedure and place it over Figure 3.4, the salinity levels for your 50 sample points can be read directly off the diagram.

On your map of salinity, using only your 50 point values, draw in isolines at intervals of 5‰ and compare this with the 'real' isoline map of salinity given in Appendix A. This map was drawn using all the data contained in Figure 3.4.

The question of which is the 'better' method to adopt of random and systematic spatial sampling is a very difficult one to answer. The choice must depend largely on the aims of the project for which the sample is required. If the aim is to make an inference from the sample about some global characteristic of the study area, for example to estimate average salinity of a salt marsh or

20	20	22	24	25	27	28	30	30	27	25	23	24	25	27	25
22	23	25	27	28	30	32	34	33	30	27	25	26	27	26	25
23	24	27	28	30	32	33	35	34	32	28	29	30	28		
25	26	28	29	32	34	35	36	34	32	31	32	31	29		
26	27	28	31	33	35	36	35	34	33	33	34	32	30		
27	28	30	32	34	36	37	36	34	34	35	35	35	30		
29	30	32	34	35	37	38	38	37	36	37	36	35	29		
31	32	33	35	36	38	40	40	39	38	37	36	33	28		
32	33	35	36	37	38	41	43	40	38	36	35	30	25		
33	35	37	38	39	41	42	43	40	37	36	33	29	24		
34	36	38	40	43	45	46	43	39	38	36	35	30	25		
34	37	39	42	44	45	47	44	42	41	38	36	30	27	23	
33	36	38	40	40	42	43	44	44	43	39	35	30	28	25	
33	35	39	41	40	38	39	42	43	40	38	37	36	31	27	
35	37	40	42	40	40	41	43	44	42	39	37	36	34	28	26
38	39	38	37	39	41	42	43	43	42	41	40	36	33	29	26
40	40	41	41	41	42	41	40	40	38	38	38	37	34	30	
39	40	41	42	43	42	41	39	37	36	36	36	36	32	29	
36	39	40	41	42	42	41	39	35	33	31	30	30	28	26	
37	38	40	42	44	41	40	37	35	32	30	29	27	25	24	
40	41	42	42	41	40	38	37	34	33	32	30	26	23		
41	42	43	42	41	40	38	36	33	30	30	29	25	23		
41	42	43	42	42	40	38	35	32	29	28	27				

Figure 3.4 Salinity levels in ‰

Stratified Samples

A geographer is studying environmental perception in Gestalton High Street. He is particularly interested in the way in which the local people's 'images' of the town centre are affected by how long they have lived in the town. He plans to carry out a sample survey to measure this influence. At the same time he is anxious to ensure that, as far as possible, his sample should be representative of the population as a whole.

He knows, from census data and other sources, that of the town's population in the 30 to 50 year age group 60% have lived in Gestalton all their lives, 20% have lived in Gestalton less than 5 years and the remaining 20% have

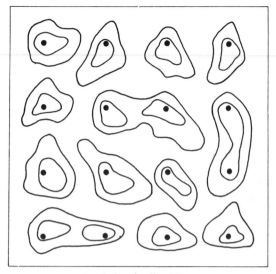

Isopleths of soil acidity

Figure 3.5 Systematic sampling: the problem of 'patterning'

lived there for between 5 and 10 years. He decides, therefore, to continue his survey until he has interviewed 120 people aged between 30 and 50 who have lived in Gestalton all their lives, 40 people in that age group who have lived there for less than 5 years and another 40 who have lived there for between 5 and 10 years.

Technically, what this geographer has done is to *stratify* his sample according to length of residence. Inhabitants in each of the three length of residence categories occur in the same proportions in the sample as they do in the total population. An alternative approach would be to select equal-sized sub-samples from each of the three length of residence groups. This form of stratification would ensure comparability between the sub-groups, but would not make the sample as a whole representative of the total population.

The geographer's sample has also been stratified according to age, in the sense that it includes only a predetermined age stratum of the town's population. This latter decision is more a question of experimental design than sampling technique. If the researcher is specifically interested in the effect of length of residence on perception he needs to 'control' as many as possible of the other variables which might affect perception. By restricting his sur-

vey to people in a fairly narrow age band he may be able to discount age as a variable affecting perception differences. Obviously there are many other variables that could be controlled for in this way, such as income, social class, sex, frequency of visiting the town centre and so on.

A general criticism that can be made of much geographical work is that geographers have not paid enough attention to experimental design, and consequently to sample design. Unlike the natural scientist, who can physically control environmental conditions in laboratory experiments, the geographer can usually only exercise effective control at the sampling stage. If a 'blanket' sample is taken and all sorts of variables are left free to vary from individual to individual, separating the effects of each of them can be a very difficult, often impossible, task. In the case of the hypothetical perception exercise, it would be extremely difficult to disentangle the effects of age, sex, length of residence, income and social class from each other. If an attempt is made to control some of these variables after the sample data has been collected, by looking at the differences in perception of individuals who are similar in all measured characteristics except one, the researcher is likely to end up working with sub-samples which contain very small numbers of individuals indeed.

Stratification can also be useful in spatial sampling. A geographer investigating recreation patterns in an urban area decides to interview 5% of the households within the city. He uses the electoral register (the list of all people entitled to vote) as a *sampling frame*, i.e. as a basis for selecting the sample. It would be possible to use random numbers to pick a 5% sample from the whole of the register, but instead, the researcher prefers to select 5% of the people in each ward (electoral sub-division) of the city. The sample obtained in this way is a 5% random sample of the city's voting population, stratified by wards. The advantage of this form of spatial stratification is that it ensures an equal degree of representation of sub-areas of the city, producing in effect a more even spatial coverage without completely sacrificing the randomness necessary for subsequent statistical testing. The use of a global random sample could result in some areas of the city being seriously underrepresented so that important local elements of the overall pattern of recreation behaviour could be overlooked. If planning decisions were subsequently made on the basis of such a survey, this local underrepresentation could have serious consequences.

Exercise 3.5

(a) Stratification of the type suggested in the last example does produce a sample which technically violates the two criteria of randomness. How?

(b) In what way(s) could the use of the electoral register bias the results of a survey concerned with studying the recreation patterns of households?

3.2 Estimates from Samples

A national sample survey of 10,000 people reveals that the average consumption of beer is 5 pints per week. The National Union of Students, conducting its own sample survey of 2,000 sixth-formers and college and university students, finds that average beer consumption amongst students is only $3\frac{1}{2}$ pints per week. Encouraged by this evidence of student sobriety, the NUS is eager to publish the results. Its statistical adviser, however, wants to know just how reliable the two estimates of average beer consumption are.

At first sight this may seem an impossible task. Common sense suggests that the larger a sample is in relation to the population from which it has been taken, the more reliable any estimates based on that sample will be. On this basis the NUS survey is likely to be more reliable than the national one, since the student sample contains a far higher proportion of all students than the proportion of the total population contained in the other sample.

Beyond this, all that is known for certain are the measured characteristics of the two samples, such as the mean beer consumption and the standard deviation of the sample values. At this stage it is necessary to make the distinction between the *sample mean* (denoted by $\bar{x}$), which is the mean as calculated from the sample values, and the *population mean* ($\breve{X}$), which is the mean for the total population. In this case the

population mean is unknown, and all that is possible is to make an estimate of the population mean from the sample values. A similar distinction exists between sample standard deviation (s) and population standard deviation (σ).

With the aid of certain assumptions, the two sample values can be used to make an estimate of the limits within which the means and standard deviations of the two populations (students and the country as a whole) can be expected to lie.

The assumptions which need to be made are: first that the samples are **random** samples, and secondly that the frequency distribution of the variable, in this case beer consumption, in the populations from which the samples are taken is **normal**. The reader can remind himself of the basic characteristics of a normal distribution by looking at Figure 3.6. The underlying logic of these assumptions will be discussed later. First the way in which the assumptions enable one to make estimates of the 'true' (population) mean and standard deviation will be discussed with reference to the two hypothetical beer drinking surveys.

Estimating the Mean

Suppose the information obtained from the two samples is as follows:

National sample—size = 10,000

mean = 5.0 pints/week

standard deviation = 2.0 pints/week

Student sample—size = 2,000

mean = 3.5 pints/week

standard deviation = 1.5 pints/week

A measure known as the *standard error of the mean* can now be calculated from this information:

$$SE_{\bar{x}} = \frac{s}{\sqrt{n}} \qquad (3.1)$$

where $SE_{\bar{x}}$ is the standard error of the mean, s is the sample standard deviation, and n is the size of the sample.

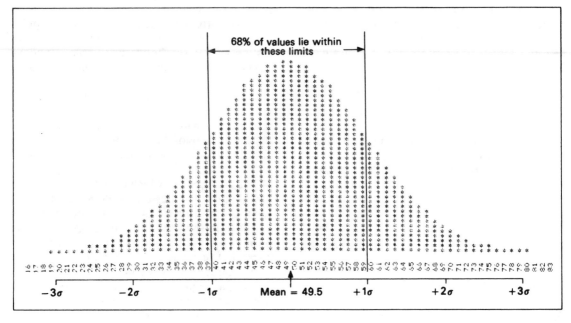

Figure 3.6 The normal distribution

For the national sample:

$$SE_{\bar{x}} = \frac{2}{\sqrt{10,000}} = \frac{2}{100} = 0.02$$

and for the student sample:

$$SE_{\bar{x}} = \frac{1.5}{\sqrt{2000}} = \frac{1.5}{44.72} = 0.03$$

The equation for the standard error of the mean is contrived in such a way that:

(1) there is a 0.682 probability that the population mean lies within the limits set by one standard error of the mean either side of the sample mean. In the case of the national beer drinkers' sample this implies that there is a 0.682 probability that the population mean is somewhere between 4.98 pints/week (5.0 − 0.02) and 5.02 pints/week (5.0 + 0.02),

(2) there is a 0.954 probability that the population mean lies within the limits set by twice the standard error of the mean on either side of the sample mean. For the national sample the population mean lies somewhere between 4.96 pints/week (5.0 − 2 × 0.02) and 5.04 pints/week (5.0 + 2 × 0.02) with a probability of 0.954,

(3) there is a 0.997 probability that the population mean lies within the limits set by three times the standard error of the mean. Based on the national sample the population mean lies somewhere between 4.94 pints/week and 5.06 pints/week with 0.997 probability.

These three relationships can be expressed in shorthand form as follows:

$$\bar{X} = \bar{x} \pm SE_{\bar{x}} \text{ with 0.682 probability} \qquad (3.2a)$$

$$\bar{X} = \bar{x} \pm 2SE_{\bar{x}} \text{ with 0.954 probability} \qquad (3.2b)$$

$$\bar{X} = \bar{x} \pm 3SE_{\bar{x}} \text{ with 0.997 probability} \qquad (3.2c)$$

where the symbol ± means 'plus or minus'.

The reader may remember that the probabilities 0.682, 0.954 and 0.997 occurred in connection with the normal distribution (Section 1.3). There is a probability of 0.682 that an individual value chosen at random from a normal distribution will be between one standard deviation less than the mean and one standard deviation more than the mean. In other words, 68.2% of all the values of a normally distributed variable can be expected to lie between the limits set by one standard deviation above and below the mean ($\bar{x} \pm \sigma$). Similarly, 95.4% of the values should be between the limits set by two standard deviations ($\bar{x} \pm 2\sigma$), and 99.7% should be within three standard deviations ($\bar{x} \pm 3\sigma$). The standard error of the mean is very closely related to standard deviation.

Exercise 3.6

Within what limits can the student population mean be expected to lie on the basis of the NUS beer drinking survey:
  (a) with 0.682 probability?
  (b) with 0.954 probability?
  (c) with 0.997 probability?

Earlier in this section it was suggested that the larger the **relative** size of the sample (i.e. its size as a proportion of the size of the population), the more reliable the sample estimates will be. The reader may have noticed, however, that in calculating the standard error of the mean only the **absolute** size of the sample is considered. In fact, as far as classical inferential statistics are concerned a large sample is inherently more reliable than a small sample for the purpose of estimating population characteristics, regardless of the size of the population. This may seem to be where statistics and common sense part company, but it is particularly valuable in the many situations in which the population under study is almost infinite. Consider the problem of estimating average sand grain size from a sample of beach sand, for example. No matter how large, the sample could not possibly contain more than an infinitessimal proportion of the total population of sand grains.

Estimating the Standard Deviation

Confidence limits can be set around sample estimates of population standard deviation, in a very similar way to that described above for the mean. An appropriate measure is the *standard error of the standard deviation*. The equation for this is:

$$SE_s = \frac{s}{\sqrt{2n}} \qquad (3.3)$$

where SE_s is the standard error of the standard

deviation, and n is the size of the sample. Applying this formula to the data obtained in the national beer-drinking survey gives a value of approximately 0.014 pints for the standard error of the standard deviation:

$$SE_s = \frac{s}{\sqrt{2n}} = \frac{2}{\sqrt{2 \times 10,000}} = \frac{2}{141.4} = 0.014$$

The standard error of the standard deviation is used in a very similar way to the standard error of the mean, by taking advantage of the following rules:

(1) $\sigma = s \pm SE_s$ with 0.682 probability

$$\qquad (3.4a)$$

(2) $\sigma = s \pm 2SE_s$ with 0.954 probability

$$\qquad (3.4b)$$

(3) $\sigma = s \pm 3SE_s$ with 0.997 probability

$$\qquad (3.4c)$$

where σ is the population standard deviation, s is the sample standard deviation, and SE_s is the standard error of the standard deviation.

In the case of the national survey of beer drinkers, therefore, it can be said that the standard deviation of beer consumption of the population is somewhere between 1.986 (2 − 0.014) and 2.014 (2 + 0.014) pints per week with 0.682 probability. There is a probability of 0.954 that the population standard deviation lies between the limits set by 1.972 (2 − 2 × 0.014) and 2.028 (2 + 2 × 0.014) pints per week, and a probability of 0.997 that the population standard deviation is between 1.958 (2 − 3 × 0.014) and 2.042 (2 + 3 × 0.014) pints per week.

Exercise 3.7

At the 0.997 probability level, what are the limits within which the standard deviation for the student population can be expected to lie, on the basis of the NUS survey?

Once again it should be noted that it is the absolute size of the sample which is used in calculating the standard error of the standard deviation, not size relative to the size of the population.

Estimates from Small Samples

The reader should by now have realized that a statistician is a cautious person. When making estimates about populations on the basis of very small samples the statistician tends to be even more cautious. In the preceding sections the sample standard deviation has been used directly in calculating the confidence limits within which the population mean and population standard deviation can be expected to lie. When working with very small samples, and certainly when the sample size is 10 or less, it is necessary to adopt wider confidence limits.

The sample standard deviation as an estimate of the population standard deviation is replaced by the *best estimate of the standard deviation,* denoted by the symbol $\hat{\sigma}$ ('sigma cap'). This can be calculated from the sample standard deviation as follows:

$$\hat{\sigma} = s\sqrt{\frac{n}{n-1}} \qquad (3.5)$$

where s is the sample standard deviation, and n is the sample size. Alternatively, the best estimate of the standard deviation can be calculated directly from the sample data by using a modified form of the equation for the standard deviation:

$$\hat{\sigma} = \sqrt{\frac{\sum (x - \bar{x})^2}{n-1}} \qquad (3.6)$$

or:

$$\hat{\sigma} = \sqrt{\frac{\sum x^2 - n\bar{x}^2}{n-1}} \qquad (3.7)$$

These three equations should give identical values when applied to the same set of data, the choice of which to use is merely a matter of convenience. In most cases it is probably fastest to use Equation 3.7. It is also likely to be more accurate than Equation 3.5 which involves taking square roots twice: once in the initial calculation of s, and again in Equation 3.5 itself.

The difference between the sample standard deviation and the best estimate can be demonstrated with a simple hypothetical sample of values of a variable, x (Table 3.2). In this rather extreme case (a sample size as small as five is not to be recommended!) the effect of using the best estimate of the standard deviation in place of the sample standard deviation is to increase the estimate of the population standard deviation by more than 10%, from 0.895 to 1.0.

There is a similar effect on the confidence limits for the estimate of the mean. Using Equation 3.1 the standard error of the mean for this sample can be calculated as:

$$SE_{\bar{x}} = \frac{s}{\sqrt{n}} = \frac{0.895}{\sqrt{5}} = \frac{0.895}{2.236} = 0.400$$

So, at the 0.954 probability level:

$$\bar{X} = \bar{x} + 2SE_{\bar{x}} = 3 \pm 0.800$$

Table 3.2 Sample Standard Deviation and Best Estimate

x	x^2	sample standard deviation
2	4	$s = \sqrt{\dfrac{\sum x^2}{n} - \bar{x}^2} = \sqrt{\dfrac{49}{5} - 3^2}$
4	16	
3	9	
4	16	$= \sqrt{9.8 - 9} = \sqrt{0.8} = 0.895$
2	4	

$\sum x = 15 \quad \sum x^2 = 49$

$\bar{x} = \dfrac{\sum x}{n} = \dfrac{15}{5} = 3$

best estimate

$$\hat{\sigma} = \sqrt{\frac{\sum x^2 - n\bar{x}^2}{n-1}} = \sqrt{\frac{49 - (5 \times 3^2)}{5-1}}$$

$$= \sqrt{\frac{49 - 45}{4}} = \sqrt{\frac{4}{4}} = 1.000$$

i.e. the population mean can be expected to lie between 2.200 and 3.800.

For small samples, a different equation for the standard error of the mean is needed, substituting the best estimate of the standard deviation for the sample standard deviation:

$$SE_{\bar{x}} = \frac{\hat{\sigma}}{\sqrt{n}} \qquad (3.8)$$

Applying this to the example data given:

$$SE_{\bar{x}} = \frac{1}{\sqrt{5}} = \frac{1}{2.236} = 0.447$$

The confidence limits at the 0.954 level can now be recalculated as:

$$\bar{X} = 3 \pm 0.894 \text{ (i.e. between 2.106 and 3.894)}$$

The result of using the best estimate in place of the sample standard deviation is always to spread further apart the limits within which the population mean can be expected to lie. In effect this means adopting a more cautious attitude.

A similar approach should be used in calculating the confidence limits for the population standard deviation, when working with small samples. The usual equation for the standard error of the standard deviation (equation 3.3) should be replaced by one involving the best estimate:

$$SE_s = \frac{\hat{\sigma}}{\sqrt{2n}} \qquad (3.9)$$

Exercise 3.8

Using Equation 3.9 calculate the standard error of the standard deviation of the hypothetical variable, x (Table 3.2). At the 0.954 probability level within what limits can the population standard deviation be expected to lie, based on the sample data?

The reader may well wish to know how small is small? The fact is that as sample size (n) increases it matters less and less whether sample measures or best estimates are used in the calculation of standard errors and confidence limits. As n increases the value of $\sqrt{n/(n-1)}$ becomes closer and closer to one. The closer this value is to one, the less difference there will be between the sample standard deviation (s) and the best

estimate $(\hat{\sigma})$. Certainly once sample size exceeds 20 the difference between the sample standard deviation (s) and the best estimate of the standard deviation $(\hat{\sigma})$ is negligible.

Estimating Proportions

The forms of standard error dealt with above can only be applied to estimates of the mean and the standard deviation of interval measurements. In some cases, however, the purpose of a sampling exercise may be to estimate the proportion of a population which possesses a certain attribute. In Exercise 3.1 for example the aim is to predict what percentage have a television set, what percentage own a car, and what percentage would vote for the N.D.P.

There is a form of standard error which can be used to place confidence limits about these proportional estimates. This is the *standard error of a percentage estimate*:

$$SE\% = \sqrt{\frac{pq}{n}} \qquad (3.10)$$

where SE% is the symbol used here for the standard error of a percentage estimate, p is the percentage of the sample possessing the particular attribute, q is the percentage of the sample not possessing that attribute, and n is the number of individuals in the sample.

In the worked example for Exercise 3.1, a sample of 50 produced an estimate of 86% for the percentage of the population which possessed a television (p in Equation 3.10). The estimated percentage of households without a television (q) is therefore $100 - 86 = 14$. Equation 3.10 can be applied to this data:

$$SE\% = \sqrt{\frac{pq}{n}} = \sqrt{\frac{86 \times 14}{50}} = \sqrt{\frac{1204}{50}}$$

$$= \sqrt{24.08} = 4.91$$

This standard error can be used, in a very similar way to the two forms discussed earlier, to establish confidence limits about the estimate of the population percentage. It can be shown that:

$$\text{population \%} = \text{sample \%} \pm SE\%$$

$$\text{with 0.682 probability} \qquad (3.11a)$$

population % = sample % ± 2SE%

with 0.954 probability   (3.11b)

population % = sample % ± 3SE%

with 0.997 probability   (3.11c)

For the present example, this means that there is a 0.682 probability that the percentage of households in the population with a television is somewhere between 81.09% (86 − 4.91) and 90.91% (86 + 4.91). At the 0.954 confidence level the population percentage can be expected to be somewhere between 76.18% (86 − 2 × 4.91) and 95.82% (86 + 2 × 4.91).

Exercise 3.9

Between what limits can the percentage of car owners in the population be expected to lie at the 0.954 confidence level, on the basis of the worked example for Exercise 3.1 (a sample percentage of 46% and a sample size of 50)?

Simulating Standard Error

The business of making estimates from samples lies at the heart of one of the most important branches of statistics, inferential statistics. It is therefore important to understand the logic of standard error before going on to the next chapters. The following paragraphs describe a simulation which may help to clarify the way in which standard error 'works', and indeed to demonstrate that it **does** 'work'.

Figure 3.6 shows an artificially constructed normal distribution with a mean of 49.5 and a standard deviation of 10. Each column of the histogram contains, to the nearest whole number, the correct frequency, as predicted by a normal distribution (Section 1.3). 50 samples each of twenty numbers were taken at random from this population, and the mean value of each sample calculated. Figure 3.7 shows the frequency distribution of these 50 sample means. The sample means range from 45.65 to 53.95, but they cluster around the population mean of 49.5.

In fact the frequency distribution of the means of a large number of random samples taken from a common, normally distributed

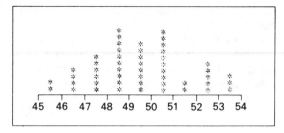

Figure 3.7 Distribution of sample means

population (the sampling distribution of the mean) is itself a normal distribution. Furthermore, when the standard error of the mean is calculated for each of the 50 samples it is found that in 33 cases out of 50 (a probability of 0.66) the population mean does lie within the limits set by plus or minus one standard error of the sample mean. In 49 cases out of 50 (0.98 probability) the population mean was within the limits set by plus or minus two standard errors.

These probabilities are quite close to the theoretical probabilities of 0.682 and 0.954, particularly in view of the fact that the experiment, or simulation, involved taking only 50 samples. A more extensive simulation should produce probabilities even closer to the theoretical values.

Like the simulation described in Section 1.4, this one was carried out by computer. It could, however, be carried out as a class exercise as follows:

(1) Each student picks a different row of numbers from a table of normally distributed random numbers (Appendix D2). This gives a set of random samples from a normally distributed population, one sample for each student.

(2) Each student calculates the mean, standard deviation and standard error of the mean for his sample.

(3) The mean of the total population is estimated. This can be done by taking the mean of all the sample means. Because of the way in which the numbers in Appendix D2 were generated, the mean should be very close to zero.

(4) A histogram of the distribution of sample means (like Figure 3.7) is made.

(5) Each student calculates whether the population mean is contained within the limits set by (a) one, (b) two, and (c) three standard errors of his sample mean.

(6) The percentage of samples falling into each of the categories (a), (b) and (c) above is calculated. These can be compared with the 'true' percentages of 68.2% and, 95.4% and 99.7% respectively.

All things being equal, the correspondence between the simulated and theoretical percentages will improve as more samples are taken and as sample size is increased. Both these assertions can be tested by modifying the simulation.

3.3 Sample Size

In Section 3.2 various measures of standard error were introduced. These enable the statistician to place confidence limits around estimates made from samples, on the basis of certain assumptions (normally distributed population and random samples). In order to do this, only two things need to be known about the sample, its size and the standard deviation of the values in it. It is a fairly straightforward piece of algebra to turn the equation for the standard deviation around so that it can be used to calculate the minimum sample size necessary to estimate the population mean, or standard deviation, within specified confidence limits.

The reader may recall that the equation for the standard error of the mean is as follows:

$$SE_{\bar{x}} = \frac{s}{\sqrt{n}} \qquad (3.1)$$

If a desired value for the standard error of the mean is specified in advance, Equation 3.1 can be rearranged to calculate n:

$$SE_{\bar{x}} = \frac{s}{\sqrt{n}} \qquad (3.1)$$

$$\therefore SE_{\bar{x}}\sqrt{n} = s$$

$$\therefore \sqrt{n} = \frac{s}{SE_{\bar{x}}}$$

$$\therefore n = \left(\frac{s}{SE_{\bar{x}}}\right)^2 \qquad (3.12)$$

The way in which this new equation can be used can be demonstrated with reference to the hypothetical national beer drinking survey

(Section 3.2). Based on a sample of 10,000, this survey produced an estimate of average beer consumption in the national population of 5 pints per week. The sample standard deviation was 2 pints, and the standard error of the mean 0.02 pints. This low standard error of the mean enabled very narrow confidence limits to be placed around the estimate of the population mean. In fact there is a probability of 0.682 that the population mean lies somewhere between 4.98 and 5.02 pints per week (5 ± 0.02).

Suppose, however, the organizers of this survey had been content to estimate the mean national beer consumption just to the nearest 1/4 pint at the 0.682 probability level: how big need the sample have been? From Equation 3.12 the minimum necessary sample size can be calculated, using 0.25 as the desired value of $SE_{\bar{x}}$:

$$n = \left(\frac{s}{SE_{\bar{x}}}\right)^2 = \left(\frac{2}{0.25}\right)^2 = 8^2 = 64$$

A sample size of only 64 would have sufficed to give an estimate of the national (population) mean which was accurate to within plus or minus 1/4 pint at the 0.682 probability level.

Suppose the same degree of accuracy were required, but at the 0.954 probability level. Confidence limits at this level are set by plus or minus two standard errors of the mean (Equation 3.2b). This means that $2 \times SE_{\bar{x}}$ needs to be 0.25, i.e. that $SE_{\bar{x}}$ needs to be $0.25/2 = 0.125$. Equation 3.12 can now be used as before:

$$n = \left(\frac{s}{SE_{\bar{x}}}\right)^2 = \left(\frac{2}{0.125}\right)^2 = 16^2 = 256$$

To obtain an estimate of average national beer consumption that was accurate to the nearest 1/4 pint at the 0.954 probability level, a sample size of at least 256 would be necessary.

Exercise 3.10

The hypothetical NUS survey of 2,000 students produced an estimate of average beer consumption of $3\frac{1}{2}$ pints per week. The sample standard deviation was $1\frac{1}{2}$ pints per week. What sample size would be necessary to produce an estimate of the mean that was accurate to plus or minus $\frac{1}{2}$ pint at the 0.997 probability level?

A similar technique can be used to calculate minimum sample size for surveys concerned with estimating percentages. The equation for the percentage standard error is:

$$SE\% = \sqrt{\frac{pq}{n}} \qquad (3.10)$$

$$\therefore (SE\%)^2 = \frac{pq}{n}$$

$$\therefore n(SE\%)^2 = pq$$

$$\therefore n = \frac{pq}{(SE\%)^2} \qquad (3.13)$$

By setting SE% to a desired value, Equation 3.13 can be used to calculate minimum sample size. This can be demonstrated in connection with the Spoton survey (Exercise 3.1). The worked example, based on sampling every fourth house in the list, produced an estimate of the proportion of households owning televisions of 86%. Sample size was 50. Suppose an estimate of the television ownership level which was accurate to within plus or minus 10% at the 0.954 probability level is desired. The 0.954 probability confidence limits are set by $2 \times$ SE%, and this needs to be 10%. The standard error therefore needs to be set at 5% (10/2). Equation 3.13 can now be applied:

$$n = \frac{pq}{(SE\%)^2} = \frac{86 \times 14}{(5)^2}$$

$$= \frac{1204}{25} = 48.2$$

Minimum sample size to achieve this degree of accuracy would therefore be 49.

Exercise 3.11

(a) The worked example for Exercise 3.1 produced an estimate of N.D.P. voting of 34%, with a sample size of 50. What minimum size of sample would be necessary to ensure an estimate of N.D.P. voting which was accurate to within plus or minus 15% at the 0.997 probability level?

(b) (Class exercise) Calculate minimum sample size necessary to estimate % N.D.P. voters to the nearest 10% at the 0.954 probability level on the basis of your own sampling results.

The reader may find that the methods discussed in this section can sometimes produce estimates of minimum sample size larger than the population. This paradox is one of the penalties of methods which rely on absolute rather than relative sample size. However, it has already been pointed out that such methods are absolutely essential for dealing with the very common situation in which population size is uncountably large.

4 Comparisons

Many geographers would agree that the concept of spatial differentiation lies close to the heart of their subject. Much research effort is devoted to the study of differences between areas in terms of the numerous phenomena that geographers are interested in. Biogeographers, for example, might look for differences in vegetation between areas with contrasting soil, slope or aspect characteristics. Urban geographers investigate differences in the industrial structure or socio-economic patterns of spatially distinct parts of an urban area. In a similar way geomorphologists might investigate differences in slope angle between valley sides of different lithology or aspect.

In each case the ultimate aim of the geographer is probably to explain apparent differences. Before looking for possible explanations, however, he will want to know how large is the difference and also how significant it is. Comparative statistics can help to answer these two questions. In the first instance they provide a descriptive measure of difference between sets of data. When the data relate to sample measurements, descriptive statistics also enable inferences to be made about differences between the populations from which the samples have been taken. Statistical inference has already been discussed at some length in Section 1.4.

In this chapter comparative statistics will be discussed for dealing with three types of situation: comparisons between one set of data and a theoretical frequency distribution, comparisons between two sets of data, and comparisons between three or more sets of data. The appropriateness of any particular technique depends also on the scale of measurement (nominal, ordinal, or interval) of the data, and on the assumptions inherent in the technique. Some of the methods discussed in this chapter can only be used to make inferences from sample data if it can be assumed that the sample measurements have been taken at random from populations in which the particular measured characteristic is normally distributed. Such techniques are referred to as *parametric* techniques. In the case of *nonparametric* techniques no assumptions are made about the frequency distribution of the data. The distinction between parametric and nonparametric methods is an important one. If the reader is not clear on this point he should refer back to Section 1.4.

4.1 The Kolmogorov-Smirnov Test

In previous sections mention has been made of different types of theoretical frequency distribution, such as the Poisson and the normal. In many geographical situations it may be of interest to know whether it is likely that a sample of data has been drawn from a population which has a particular type of distribution.

A geographer travelling through a rural area develops a hypothesis that the distribution of garages along the road is random. To test this hypothesis he keeps a record for 100 miles of the number of garages passed in each mile stretch of road. Table 4.1 shows the data he collects expressed as a frequency distribution (column b). In column c the frequencies have

Table 4.1 Observed Distribution of Garages and Theoretical Poisson Distribution

a Number of garages	b Frequency	c Probabilities observed	d Poisson	e Cumulative probabilities observed	f Poisson	g difference
0	30	0.30	0.13	0.30	0.13	0.17
1	25	0.25	0.26	0.55	0.39	0.16
2	15	0.15	0.27	0.70	0.66	0.04
3	12	0.12	0.19	0.82	0.85	0.03
4	5	0.05	0.10	0.87	0.95	0.08
5	6	0.06	0.04	0.93	0.99	0.06
6	2	0.02	0.01	0.95	1.00	0.05
7	1	0.01	0.00	0.96	1.00	0.04
8	0	0.00	0.00	0.96	1.00	0.04
9	0	0.00	0.00	0.96	1.00	0.04
10	4	0.04	0.00	1.00	1.00	0.00
	100	1.00	1.00			

been converted into probabilities by dividing each one by 100 (the total number of mile stretches of road). In order to test the hypothesis of a random distribution, this observed probability distribution is compared with a theoretical Poisson probability distribution. The Poisson distribution is appropriate in this case as the hypothesis is that the garages occur at random in a continuum of space (Section 1.3). Poisson probabilities corresponding to the observed probabilities are given in column d of Table 4.1.

What is now needed is a test of how closely the observed probabilities correspond to the theoretical probabilities, a test of 'goodness-of-fit'. The Kolmogorov–Smirnov test can perform this function.

To compare two probability distributions using the Kolmogorov–Smirnov test, the distributions have to be converted into cumulative probability distributions. Columns e and f of Table 4.1 show these cumulative probabilities; for example, column e simply keeps a running total of the values in column c by summing from the top downwards. These two cumulative probability distributions can be visualized as two step-like graphs, as shown in Figure 4.1.

The *null hypothesis* for the Kolmogorov–Smirnov goodness of fit test is that the sample data which produced the observed probability

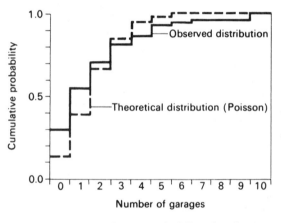

Figure 4.1 Cumulative probability distributions for Kolmogorov–Smirnov test

distribution have been drawn from a population which possesses the specified theoretical distribution, in this case the Poisson distribution.

The Kolmogorov–Smirnov statistic is simply the maximum absolute difference between the theoretical and the observed cumulative probability distributions, written as D. Column g in Table 4.1 gives the absolute difference between columns e and f. The largest value in column g is 0.17, and this is therefore the value of the Kolmogorov–Smirnov statistic, D. Since the sampling distribution of D under the null hypothesis is known, tables of critical values of D are available (Appendix C1).

The degrees of freedom for the Kolmogorov–Smirnov goodness of fit test are the number of items in the observed frequency distribution, or the sum of the observed frequencies (Table 4.1a). In this case the degrees of freedom are therefore 100, i.e. the total number of one-mile stretches of road.

If the calculated value of D is greater than the tabled critical value at a specified significance level, the null hypothesis can be rejected at that level. If a significance level of 0.05 is decided upon in the present example, with 100 degrees of freedom the critical value of D obtained from Appendix C1 is 0.14.

Since the calculated value of D is 0.17, which is greater than the critical value, the null hypothesis can be rejected at the 0.05 significance level. In effect this means that there is a probability of less than 0.05, or 5 chances in 100, that the sample data have come from a population which possesses a Poisson distribution. The geographer can conclude that the distribution of garages in this rural area is very unlikely to be random.

Exercise 4.1

Using the data on children in care given in Table 1.1, and the expected probabilities predicted by a normal distribution (produced as part of the answer to Exercise 1.1), calculate the value of the Kolmogorov–Smirnov statistic, D. You will first need to convert the two probability distributions into two cumulative probability distributions. Is the observed distribution significantly different from a normal distribution at the 0.05 level?

In doing Exercise 4.1, the reader may have realized that the number of classes used and the choice of class boundaries in creating the observed frequency distribution could affect the calculation of the Kolmogorov–Smirnov statistic, D. By using fewer classes it is often possible to decrease the maximum divergence between the observed and theoretical probability distributions. In general it is desirable to put the data into as many classes as possible before applying the Kolmogorov–Smirnov test, in

order to avoid obscuring major divergences between the observed and theoretical distributions.

The chi square test can also be used as a test of goodness-of-fit. However, the Kolmogorov–Smirnov test is in many ways easier to use and need not be hampered by some of the restrictions which apply to the chi square test, discussed in Section 4.6 and elsewhere (for example Siegel (1956) p. 51).

4.2 A Runs Test for Randomness

The Poisson distribution can be used to calculate the probability of any particular frequency of occurrence of an 'event' which occurs at random in a continuum of space or time (see Section 1.3). However, the Poisson distribution cannot really be used as a test of randomness in the order of occurrence of events, either in time or in space.

It has been suggested in Section 1.3 that thunderstorms can be considered to be random events. The probability of having any specified number of thunderstorms during a set time period can therefore be predicted by the Poisson distribution. Table 4.2 contains a frequency distribution of thunderstorms per year, based on data for a 50 year period, and corresponding closely to a Poisson distribution. However,

Table 4.2 Frequency Distribution of Thunderstorms per Year

Number of thunderstorms per year	Number of years with this many thunderstorms
0	1
1	2
2	4
3	7
4	9
5	9
6	7
7	5
8	3
9	2
10	1

Table 4.3 Possible Sequences of 5 Thundery and 5 Non-Thundery Years

	Sequence	Runs	
a	T T T T T N N N N N	2	T = thundery year
b	T N T N T N T N T N		N = non-thundery year
c	T T T T N N T N N N	4	runs are underlined in a and c
d	N N T N N T N T T T		

the frequency distribution gives no information about the order in which particular types of years occurred. It could be that there was a regular increase in the number of thunderstorms per year throughout the 50 year period. Alternatively, each 'thundery' year could have been followed by a 'non-thundery' year. If the years were classified as either 'thundery' or 'non-thundery', perhaps on the basis of whether or not there had been more than five thunderstorms during the year, a large number of different orderings would be possible. Table 4.3 shows a number of alternative sequences of 5 thundery and 5 non-thundery years. The two sequences shown in Table 4.3a and b may seem intuitively too 'regular' to be the result of a random process. Table 4.3c and d shows two more sequences, which may appear intuitively to be 'more random'.

The runs test can be used to estimate the probability that a particular sequence of alternatives, such as thundery and non-thunder years, could have been produced by chance. The ordering is measured by counting the number of *runs* in the sequence, a run being a group of one or more like elements. In Table 4.3a there are two runs, and in Table 4.3c four runs.

Exercise 4.2

How many runs are there in the sequences shown in Table 4.3b and d?

Significance

Given a particular number of elements of each of two alternative types there is a finite number of ways in which they can be ordered. For example, if there are two thundery and three

Table 4.4 All Possible Sequences of 2 Thundery and 3 Non-Thundery Years

Sequence	Number of runs
T T N N N	2
T N T N N	4
T N N T N	4
T N N N T	3
N T T N N	3
N T N T N	5
N T N N T	4
N N T T N	3
N N T N T	4
N N N T T	2

non-thundery years in a sequence of five years, there are 10 possible orderings. These are shown in Table 4.4, together with the number of runs produced by each ordering. If the orderings are produced by a random process, such as drawing cards marked with T or N from a hat, each of these orderings has the same probability of occurrence. However, it can be seen that, whereas only one of these orderings produces as many as five runs, there are four possibilities of getting four runs. It is possible to calculate the probability of getting any particular number of runs in a sequence containing two Ts and three Ns. Two runs can occur in two out of ten alternative arrangements, giving a probability of 0.2. The other probabilities can be worked out in a similar way:

3 runs occur in 3 out of 10 ways
∴ probability = 0.3

4 runs occur in 4 out of 10 ways
∴ probability = 0.4

5 runs occur in 1 out of 10 ways
∴ probability = 0.1

The probability distribution of the number of runs in this situation is shown in Figure 4.2.

It can be seen that in this example, and in fact in all cases where only two alternatives are involved, there is a relatively high probability of getting a medium number of runs and a relatively low probability of getting a small or a large number of runs. In order to test whether a particular sequence is significantly different from a random one it is necessary to decide whether the number of runs observed in the sequence is either so large or so small that it is very unlikely to have come about by chance.

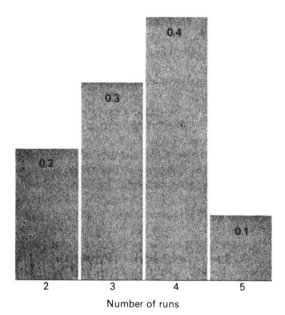

Figure 4.2 Probability distribution of the number of runs when $n_1 = 2$ and $n_2 = 3$

It would be possible to calculate the probability of getting any particular number of runs given a specified number of elements of each of two alternative types. Fortunately tables are available giving critical values of the number of runs (Appendix C2). For any pair of values of n_1 (the number of elements of one type in a sequence) and n_2 (the number of elements of the other type) two critical values are tabled. With $n_1 = 6$ and $n_2 = 9$, for example, the critical values at the 0.05 significance level are 4 and 13. This means that there is a probability of 0.05 of obtaining 4 or fewer or 13 or more runs in a random sequence containing 6 elements of one

type and 9 elements of the other type. A sequence of this sort which contains 4 or less runs, or 13 or more runs can be said to depart from random ordering at the 0.05 significance level.

Exercise 4.3

(a) Draw up a table showing all the possible orderings of alternatives for $n_1 = 3$, $n_2 = 3$.

(b) What is the probability of getting only 2 runs?

(c) What is the probability of getting 5 or more runs?

(d) Table 4.5 shows the number of cars arriving at a country park on a summer Sunday during half-hour periods between 9.00 and 20.30. Each half-hour period can be classified as 'busy' or 'slack' according to whether the number of cars arriving during that period is greater than or less than the median for the whole day. Use the runs test to determine whether the observed sequence of 'busy' and 'slack' periods is different from a random one at the 0.05 significance level.

Table 4.5 Arrival of Cars at a Country Park

Time	Arrivals	Time	Arrivals
9.00	4	15.00	32
9.30	12	15.30	30
10.00	41	16.00	20
10.30	52	16.30	15
11.00	55	17.00	9
11.30	40	17.30	12
12.00	38	18.00	6
12.30	43	18.30	5
13.00	21	19.00	20
13.30	16	19.30	7
14.00	18	20.00	9
14.30	14	20.30	2

4.3 The Mann–Whitney U test

The Mann–Whitney U test is a fairly simple test of whether there is a significant difference between **two** sample sets of data. It is a **non-parametric** test, which means that it is not restricted by any assumptions about the nature of the population(s) from which the samples have been taken (see Section 1.4). It is applicable to **ordinal** (ranked) data. Since interval data can easily be converted into ordinal form (Section 1.2), the Mann–Whitney U test provides a valuable alternative to the Student's t test (Section 4.4). In many cases when comparing samples of geographical data assumptions of normality are quite unrealistic (see Section 2.3). In such circumstances use of the Student's t test is invalid, and a nonparametric alternative such as the Mann–Whitney U test should always be used.

The Mann–Whitney U test is a test of the significance of a difference between two samples. The null hypothesis is that the two samples are taken from a common population, so that there should be no consistent difference between the two sets of values. Any observed difference between the two samples, such that one set of values is consistently larger than the other is due entirely to chance in the sampling process.

A simple example will show how the Mann–Whitney U test is applied. Sample measurements of soil moisture content are taken at randomly located points in two areas of woodland. These two sets of values are given in Table 4.6 (columns x and y). To apply the U test, the data must first be ranked from lowest to highest, and these ranks are given in Table 4.6, columns

r_x and r_y. The reader should note that the values have been ranked **out of the total set of data**, i.e. they are the **overall** rankings of the values not the rankings within the samples. It should also be noted that identical values are given the arithmetic average of the rankings they would otherwise receive. For example the value 9.3 occurs twice in the data given in Table 4.6. These values occupy the 4th and 5th positions in the overall rankings, so they are each given the ranking 4.5 $((4+5)/2)$. If there were three identical values occupying 5th, 6th and 7th places, these would each be given the ranking 6 $((5+6+7)/3)$. The sum of these rankings is then calculated for each sample ($\sum r_x$ and $\sum r_y$ in Table 4.6) and used in calculating U from the following equations:

$$U_x = n_x n_y + \frac{n_x(n_x + 1)}{2} - \Sigma r_x \qquad (4.1)$$

$$U_y = n_x n_y + \frac{n_y(n_y + 1)}{2} - \Sigma r_y \qquad (4.2)$$

Where n_x and n_y are the number of individuals in sample x and sample y respectively.

Using Equation 4.1 the value of U_x is:

$$6 \times 5 + \frac{6(6 + 1)}{2} - 44.5$$

$$= 30 + \frac{42}{2} - 44.5 = 51 - 44.5 = 6.5$$

Table 4.6 Calculation U for the Mann–Whitney Test

x	r_x	y	r_y	
15.2	11	8.4	1	
10.7	8	9.3	4.5	
8.6	2	8.7	3	$n_x = 6$
9.3	4.5	10.2	7	$n_y = 5$
12.4	10	10.0	6	
11.1	9			
	$\sum r_x = 44.5$		$\sum r_y = 21.5$	

$$U_x = n_x n_y + \frac{n_x(n_x + 1)}{2} - \Sigma r_x = 30 + \frac{42}{2} - 44.5 = 6.5$$

$$U_y = n_x n_y + \frac{n_y(n_y + 1)}{2} - \Sigma r_y = 30 + \frac{30}{2} - 21.5 = 23.5$$

From Equation 4.2 the value of U_y is:

$$6 \times 5 + \frac{5(5 + 1)}{2} - 21.5$$

$$= 30 + \frac{30}{2} - 21.5 = 45 - 21.5 = 23.5$$

As a check on the calculations it can be noted that $U_x + U_y = n_x \times n_y$. In this case $n_x \times n_y = 5 \times 6 = 30$, and $U_x + U_y = 6.5 + 23.5 = 30$.

The value of U needed for the significance test depends on the form of hypothesis that is being tested. This will be explained in the next two sections.

Significance

Tables of critical values of U are available (Appendix C3) and can be consulted to find out whether a particular observed difference between two samples can be considered significant at particular probability levels. In fact it is possible to derive the sampling distribution of U directly, without recourse to statistical theory. A simple example will show how this can be done.

If there are two samples, x and y, containing two and three individuals respectively ($n_x = 2$, $n_y = 3$), there are 10 different ways in which the overall rankings could be split between the two samples (ignoring the possibility of tied ranks). These ten different ways are shown in Table 4.7, together with the value of U (the smaller of the two possible values) produced by each one. It should be noted that a **low** value of U is produced when there is a **large** difference between the two samples. The first and last combinations shown in Table 4.7 both produce U values of 0. This indicates that there is the maximum possible amount of difference between the samples, and it can be seen that there is no overlap in the rankings of the two samples in either case.

Table 4.7 (a) *All Possible Combinations of Ranks in Two Samples of 2 and 3 Individuals in the Mann–Whitney U Test*

Ranks		Mean ranks		U (smaller of U_x and U_y)
Sample x	Sample y	x	y	
1, 2	3, 4, 5	1.5	4	0
1, 3	2, 4, 5	2	3.7	1
1, 4	2, 3, 5	2.5	3.3	2
2, 3	1, 4, 5	2.5	3.3	2
1, 5	2, 3, 4	3	3	3
2, 4	1, 3, 5	3	3	3
2, 5	1, 3, 4	3.5	2.7	2
3, 4	1, 2, 5	3.5	2.7	2
3, 5	1, 2, 4	4	2.3	1
4, 5	1, 2, 3	4.5	2	0

Mean Rank x < Mean Rank y (first four rows); Mean Rank x > Mean Rank y (last four rows)

Table 4.7 (b) *Probability Distribution of U for $n_x = 2$, $n_y = 3$*

U	Frequency	Probability	Cumulative probability
0	2	0.2	0.2
1	2	0.2	0.4
2	4	0.4	0.8
3	2	0.2	1.0
	10	1.0	

Table 4.7 also gives the mean rank for each sample of x and y values. Remember that the two sets of sample values used in the U test are ranked **together** from lowest to highest. A low mean rank for one sample indicates that it contains generally smaller values than the other sample. From Table 4.7 it can be seen that, for example, a U value of 0 can be produced when the mean rank of x is less than the mean rank of y, **or** when it is greater than the mean rank of y. The same can be said of the ways in which U values of 1 and 2 can be obtained. The sampling distribution of U is thus symmetrical, as shown in Figure 4.3. The left hand 'tail' contains low values of U produced in situations in which the mean rank of x is less than the mean rank of y. The right hand 'tail' contains low values produced when the mean rank of x is greater than the mean rank of y. This is important when considering the different forms of alternative hypothesis against which the null hypothesis can be tested.

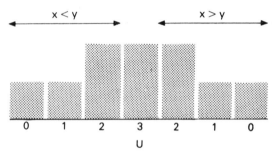

Figure 4.3 Sampling distribution of U

One-tailed and two-tailed tests

Three possible alternative hypotheses can be formulated:
(1) H_1: sample x and sample y come from populations with different mean ranks (without specifying which of the two mean ranks is believed to be the larger). This can be written in shorthand as H_1: $X \neq Y$ (using capital letters to denote populations).
(2) H_1: sample x and sample y come from populations which have different mean ranks, **and** the mean rank of population X is greater than the mean rank of population Y (H_1: $X > Y$).

(3) H_1: sample x and sample y come from populations which have different mean ranks, **and** the mean rank of X is less than the mean rank of Y (H_1: $X < Y$).

Suppose a U test is carried out on two samples, x and y, containing 2 and 3 individuals respectively, and that a U value of 0 is produced. From Table 4.7 it can be seen that a U value as low as this could come about in two ways (the first and last combinations of ranks). The overall probability of getting a U of 0 is therefore $2/10 = 0.2$. The probability that the null hypothesis true is therefore also 0.2. The null hypothesis could be rejected in favour of the first alternative hypothesis (H_1: $X \neq Y$) at the 0.2 significance level.

Of the two ways in which a U value of 0 could occur with $n_x = 2$ and $n_y = 3$ only the second one (bottom combination in Table 4.7) is compatible with the second alternative hypothesis (H_1: $X > Y$). The probability that the null hypothesis is true when set against this alternative hypothesis, which specifies a direction of difference, is therefore only $1/10 = 0.1$. The null hypothesis could be rejected in favour of the second (directional) alternative hypothesis at the 0.1 significance level.

A similar line of reasoning can be applied in connection with the third alternative hypothesis (H_1: $X < Y$). This time only the first way in which a U value of 0 could be produced (top line of Table 4.7) is compatible with the alternative hypothesis. The third alternative hypothesis could therefore be accepted at the 0.1 significance level.

A test against a non-directional alternative hypothesis ($X \neq Y$) is a *two-tailed test*. It is concerned with the probability of getting by chance a low value of U which could occur in either 'tail' (end) of the sampling distribution (Figure 4.3). A test against a directional alternative hypothesis ($X > Y$, or $X < Y$) is a *one-tailed test*. Only the probability of getting by chance a low value of U in a way that is compatible with the alternative hypothesis is considered. In other words, only one 'tail' of the sampling distribution is involved.

In passing, it might be noted that some statisticians (see for example, Lieberman, 1971) have questioned the logic of one-tailed tests. There is certainly something odd about the fact that a test against a more specific alternative hypothesis (a directional one) is more 'liberal' than a test against a more general (non-directional) alternative hypothesis. The term 'liberal' is used in the sense that in order to reject the null hypothesis against a directional alternative hypothesis (a one-tailed test) a less extreme U value needs to be obtained than for a two-tailed test. However, one-tailed tests are widely applied, and the reader certainly needs to be aware of the difference between one-tailed and two-tailed tests.

As mentioned earlier, the form of alternative hypothesis employed determines which value of U needs to be calculated for the test. For H_1: $X \neq Y$ the value needed is the **smaller** of U_x and U_y. This is a two-tailed test. For H_1: $X > Y$ (a one-tailed test) the value needed is U_x. For H_1: $X < Y$ (also a one-tailed test) the value needed is U_y.

In the example given earlier (comparing two areas in terms of soil moisture content) the alternative hypothesis would probably be directional, i.e. that the area from which sample x was taken was in general wetter than the area that produced sample . This is a one-tailed test (H_1: $X > Y$), and the appropriate value of U is U_x, which has been calculated above as 6.5. Suppose a significance level of 0.05 has been decided upon. The critical value of U at this level for a one-tailed test with $n_x = 6$ and $n_y = 5$ can be found from Appendix C3 as 5. Since the calculated value is greater than the critical value, the null hypothesis can not be rejected. Note that in the Mann–Whitney U test the null hypothesis can only be rejected if the calculated value is *less than or equal to* the critical value at the chosen significance level.

Exercise 4.4

(a) Table 4.8 gives figures for the number of children per thousand who are in the care of the local authority for a sample of 'large' towns (x) and a sample of 'small' towns (y). The two samples were picked using random numbers from the data matrix in Appendix E. Small towns were defined as those with a population of less than 200,000. Use the Mann–Whitney U test to find out whether the proportion of children in the care of the local authority is significantly greater in 'large' towns than in 'small' towns at the 0.05 level.

Table 4.8 Proportion of Children in Care in Two Samples of Towns

Large towns	x	Small towns	y
Leicester	9.2	Birkenhead	5.5
Coventry	6.5	Bolton	15.5
Cardiff	6.8	Blackpool	5.5
Birmingham	10.7	Stockport	8.0
Newcastle	14.4	West Bromwich	5.9
Wolverhampton	5.4	York	7.5
Sunderland	6.9	Warley	3.8
Teesside	7.7	Northampton	7.2
Bradford	11.9	Gateshead	10.1
Manchester	13.0	Oxford	10.5
		Solihull	2.0
		Reading	7.0
		Bournemouth	10.0
		St Helens	7.7

4.4 Student's t Test

Student's t test is a **parametric** test of the difference between **two** samples. It is applicable only to data measured on an **interval** scale. A widely used technique amongst geographers, it is often used in circumstances in which a nonparametric alternative, such as the Mann–Whitney U test, would be more appropriate.

The null hypothesis of Student's t test is that two sets of data are random samples from a common, normally distributed population, or from two identical normally distributed populations. Before applying this test the geographer should have good reason to suppose that the variable measured has a normal (or nearly normal) distribution.

Table 4.9 contains some data obtained from a small survey carried out on two housing estates. The column headed x contains the ages of a random sample of 10 heads of households from The Cedars, and column y contains the ages of a similar sample from Platt's Mead. It is assumed that the researcher has good reason to suppose that age of head of household is a normally distributed variable.

The null hypothesis is that there is no difference between the means of the populations from which the two samples were taken. In other words, if it were possible to interview every head of household on the two estates the

Table 4.9 Ages of Sample Heads of Household in Two Housing Estates

x The Cedars	y Platt's Mead	x^2	y^2
37	48	1369	2304
46	63	2116	3969
53	62	2809	3844
50	51	2500	2601
48	46	2304	2116
27	43	729	1849
65	39	4225	1521
39	21	1521	441
28	57	784	3249
42	60	1764	3600
$\sum x = 435$	$\sum y = 490$	$\sum x^2 = 20121$	$\sum y^2 = 25494$
$n_x = 10$	$n_y = 10$	$\bar{x} = \dfrac{\sum x}{n_x} = 43.5$	$\bar{y} = \dfrac{\sum y}{n_y} = 49.0$

means for the two estates would be the same. The null hypothesis assumes that the observed difference between the sample means is due to chance in the sampling process. One or both samples might contain a number of 'freak' households with old or young heads, thus distorting the means. The smaller the samples were, the easier it would be for these 'freaks' to seriously distort the comparison between them.

The alternative hypothesis is that there is a difference between the means of the two populations, a difference which is being accurately reflected in the samples. A significance level of 0.05 is decided upon.

From the data given in Table 4.9 the value of the test statistic, t, can be calculated using the following equation:

$$t = \frac{|\bar{x} - \bar{y}|}{\sqrt{\dfrac{(\sum x^2/n_x) - \bar{x}^2}{n_x - 1} + \dfrac{(\sum y^2/n_y) - \bar{y}^2}{n_y - 1}}} \quad (4.3)$$

where $\bar{x}$ and $\bar{y}$ are the means of the two samples, $|\bar{x} - \bar{y}|$ is the absolute value of the difference between the means, and n_x and n_y are the sizes of the two samples. Note that there are many alternative forms of equation 4.3, all of which should give identical values of t if applied to a common data set. Equation 4.3 is recom-

mended here as being the easiest equation to use when calculating t by 'manual' methods. It is **not** the best equation to use as the basis for a computer program to calculate t.

The calculation of the various elements of Equation 4.3 is shown in Table 4.9. The elements can then be substituted into the equation to give:

$$t = \frac{|43.5 - 49.0|}{\sqrt{\dfrac{(20121/10) - 43.5^2}{9} + \dfrac{(25494/10) - 49.0^2}{9}}}$$

$$= \frac{5.5}{\sqrt{\dfrac{2012.1 - 1892.25}{9} + \dfrac{2549.4 - 2401}{9}}}$$

$$= \frac{5.5}{\sqrt{13.316 + 16.488}} = \frac{5.5}{\sqrt{29.804}}$$

$$= \frac{5.5}{5.459} = 1.01$$

The degrees of freedom for the t test are given by a rather complicated equation:

$$df = \frac{(SX + SY)^2}{SX^2/(n_x - 1) + SY^2/(n_y - 1)}$$

where $SX = [(\Sigma x^2/n_x) - \bar{x}^2]/(n_x - 1)$ and $SY = [(\Sigma y^2/n_y) - \bar{y}^2]/(n_y - 1)$. In fact this is not too difficult to evaluate, since the values of SX and SY are calculated as part of Equation 4.3. They are the two main elements of the denominator (below the square root sign). In this case $SX = 13.316$ and $SY = 16.488$. These values can be substituted into the equation for degrees of freedom to give:

$$df = \frac{(13.316 + 16.488)^2}{13.316^2/9 + 16.488^2/9}$$

$$= \frac{29.086^2}{177.334/9 + 42.091/9}$$

$$= \frac{888.371}{49.913} = 17.8 \simeq 17$$

It will be seen that this equation for degrees of freedom will often yield a fractional number, which can be rounded down to the nearest integer. In this case the degrees of freedom are 17. From tables of the Student's t sampling distribution (Appendix C4) the critical value of t with 17 degrees of freedom for a two-tailed test at the 0.05 significance level is 2.11. Since the calculated value of t is less than the critical value, the null hypothesis can not be rejected at the 0.05 significance level.

On the basis of the t test it is safe to assume at the 0.05 significance level that there is no difference between the mean ages of the population (of heads of household) in the two estates. The apparent difference revealed by the sample means can be accounted for by chance in the sampling process.

To consider how the t test works it is necessary to look more closely at the null hypothesis. The null hypothesis is that both samples have come from a common normally distributed population. From the data given in Table 4.9 it is possible to derive two estimates of the characteristics of this population: one on the basis of sample x, the other on the basis of sample y. From sample x, the mean of this hypothesized common population is 43.5, whereas the estimate from sample y is 49.0. Using Equation 3.7 two best estimates of the population standard deviation can also be obtained. From sample x, $\hat{\sigma}_x = 11.54$, and from sample y, $\hat{\sigma}_y = 12.84$. This is all the information necessary to construct two 'pictures' of the population, and these are

shown in Figure 4.4, superimposed over each other.

Clearly there are differences, both in central tendency and dispersion between these two pictures of the population distribution. What the t test does is to enable the researcher to decide how likely it is that two pictures as different as these could be produced as a result of taking two random samples from a common, normally distributed population. Are the two 'pictures' in Figure 4.4 'impressions' of the same distribution, slightly distorted as a result of being based only on sample information, or are they impressions of two different 'originals'?

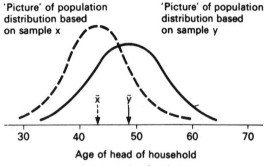

Figure 4.4 Student's t test

The denominator in Equation 4.3 is a measure known as the standard error of the difference between two means, or simply the *standard error of the difference*. Like other standard errors (see Section 3.2), the standard error of the difference is closely tied up with the normal distribution. Given two random samples from a common, normally distributed population there is a probability of 0.682 that the absolute difference between the means will not exceed one standard error of the difference. To put the same thing another way, if the absolute difference between the samples is equal to the standard error of the difference, there is a probability of 0.318 $(1 - 0.682)$ that the two samples have come from a common, normally distributed population. If the absolute difference between the means is twice the standard error of the difference, there is a probability of 0.046 $(1 - 0.954)$ that the two samples have come from a common, normally distributed population. An absolute difference between the sample means which is three times the standard error of the difference implies a probability of only 0.003 $(1 - 0.997)$ that the two samples have come from a common, normally distributed population. The probabilities mentioned are those which the reader may by now have come to associate with the normal distribution.

In the present example the absolute difference between the means is slightly greater (5.5) than the standard error of the difference (5.46). The probability that the two samples have come from a common, normally distributed population is therefore less than 0.318 but greater than 0.046.

This is not quite what the researcher wants to know. He is interested in whether the probability is less than 0.05. On the evidence of the standard error of the difference alone, this may or may not be the case. The t statistic is simply the ratio of the absolute difference between the sample means to the standard error of the difference. The advantages of using t instead of just the standard error of the difference is that it has a known sampling distribution, so that critical values of t can be specified for chosen significance levels.

One-tailed and two-tailed tests

So far attention has only focused on testing the null hypothesis against the alternative hypothesis that there is a difference between the means of the populations from which the two samples were taken. Figure 4.5 illustrates the way in which the critical value of t for this test is found. The sampling distribution of t can be thought of as a symmetrical distribution, with a value of 0 in the centre and values of t along the horizontal axis increasing outwards from the centre (Figure 4.5a). The shaded area in the left hand tail of the sampling distribution represents the probability of getting a value of t by chance under the null hypothesis as great as or greater than t_c, when $\bar{x}$ is less than $\bar{y}$. The shaded area in the right hand tail represents the same probability of getting a value of t as great as or greater than t_c when $\bar{x}$ is greater than $\bar{y}$.

For a test of the null hypothesis against the alternative hypothesis that there is a difference between the population means, the researcher is interested in both these probabilities. He therefore needs to know what area of the sampling distribution is in **both** tails of the sam-

(a)

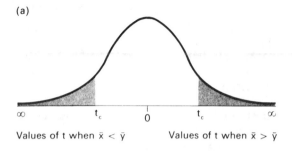

Values of t when $\bar{x} < \bar{y}$ Values of t when $\bar{x} > \bar{y}$

(b) Two-tailed test   Total shaded area
                              gives 0.05 probability

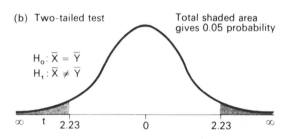

(c) One-tailed test   Shaded area gives
                              0.05 probability

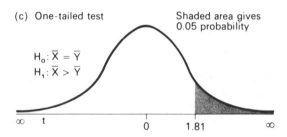

Figure 4.5 Sampling distribution of Student's t

pling distribution. Hence this form of the t test is known as a *two-tailed test*. The two-tailed critical value of t at the 0.05 probability level with 10 degrees of freedom is 2.23. This means that there is a probability of 0.05 of getting a value of t as great as or greater than 2.23 under the null hypothesis. Half of this probability (0.025) is accounted for by the area beyond 2.23 in the right hand tail of the sampling distribution. The other half is accounted for by the area beyond 2.23 in the left hand tail.

It is also possible, however, to test the null hypothesis against an alternative hypothesis that there is a difference between the population means in a specified direction. For example, the alternative hypothesis might be that the samples have come from two different populations and that the mean of population which produced sample x is greater than the mean of the population which produced sample y. In this case it is only necessary to know what area of the sampling distribution lies beyond the calculated value of t in the right hand tail. This form of the t test is a one-tailed test. The one-tailed critical value of t at the 0.05 probability level with 10 degrees of freedom is 1.81. This means that there is a probability of 0.05 of getting a value of t as great as or greater than 1.81 under the null hypothesis. This probability can be thought of as the area beyond 1.81 in one tail of the distribution. Since the distribution is symmetrical, it does not matter which tail is considered.

Figures 4.5b and c illustrate the difference between the two-tailed critical value (b) and the one-tailed critical value (c) for a t test at the 0.05 significance level with 10 degrees of freedom. From a practical point of view it is important to appreciate the difference between a one-tailed and a two-tailed test, since it determines which critical value of t is appropriate. The reader should note that in the tables of Student's t (Appendix C4) two sets of significance levels are given: one refers to a one-tailed test, and the other to a two-tailed test.

In applying a one-tailed test it is necessary to establish first of all whether the difference between the two samples is in the direction stated in the alternative hypothesis. If the alternative hypothesis is that the mean of population X is greater than the mean of population Y ($H_1: \bar{X} > \bar{Y}$) the null hypothesis can only be rejected if the mean of sample x is greater than the mean of sample y **and** if t is significant at the chosen level.

Reservations about the validity of one-tailed tests have already been expressed in Section 4.3 in connection with the Mann–Whitney U test. They apply equally to Student's t.

Exercise 4.5

Table 4.10 shows two random samples of rainfall data taken from the data matrix (Appendix E). The total rainfall for 1966 is given for 10 'northern' towns and 15 'southern' towns. 'Northern' is defined as being north of a line

Table 4.10 Rainfall Totals (1966) for Samples of Northern and Southern Towns

x Northern towns	y Southern towns
11.11	9.76
8.64	8.09
10.11	8.02
13.07	11.50
7.71	7.50
9.68	11.47
10.19	6.52
10.13	12.83
7.74	6.58
12.29	8.91
	9.50
	8.54
	7.34
	9.37
	6.30

from the Wirral to the Wash. Whether it is reasonable to expect these rainfall totals to be normally distributed in the population(s) from which the samples are taken is open to argument. For the purpose of the exercise assume that it is reasonable.

Can the null hypothesis be rejected in favour of the alternative hypothesis that 'northern' towns are wetter than 'southern' towns at the 0.01 significance level? Remember that this is a one-tailed test, since the alternative hypothesis specifies a direction of difference.

To ease the calculations, all the values have been divided by 100 (converting them from millimetres to decimetres).

4.5 The Chi Square Test

The chi square test is a very flexible test which can be applied in one-sample, two-sample and more-than-two-sample situations. To avoid repetition, all forms of the test will be covered in this section. The test is restricted to **nominal** (frequency) data and is **nonparametric**. It can be applied to other types of data, provided they are first put into nominal form (see Section 1.2). There are, however, some important limitations to the use of chi square which are sometimes ignored. These will be discussed as they arise.

The one-sample case

In the one-sample case chi square can be used as a test of the goodness of fit of an observed set of frequencies produced by a sample investigation to a theoretical frequency distribution.

A geomorphologist is studying a section of beach composed of limestone, granite and chert pebbles. Although the three parent materials are found in equal quantities in the surrounding area, the researcher has theoretical reasons for supposing that these three rock types have not contributed equal proportions of pebbles to the beach. The geomorphologist sets up a null hypothesis which states that there are equal proportions of limestone, granite and chert pebbles to be found on the beach. The alternative hypothesis is that there is in fact a difference in the number of pebbles of each type on the beach as a whole.

A random sample of 600 pebbles is collected. 180 are found to be limestone, 186 are granite and the remaining 234 are chert. Are these results compatible with the null hypothesis? According to the null hypothesis a survey of the total population of pebbles should reveal equal proportions of limestone, granite and chert pebbles. Since the data obtained refer only to a sample it is unrealistic to expect exactly equal proportions. On the other hand, proportions in the sample should not be too far from equal if the null hypothesis is to be retained. The chi square test in this situation can be used to estimate the probability that the sample of pebbles has been taken from a population in which the proportions are equal.

The chi square statistic is calculated as follows:

$$\chi^2 = \Sigma \frac{d^2}{e} \qquad (4.4)$$

where χ^2 is the symbol for chi square, d is the difference between the observed and the expected frequency for each category, and e is the expected frequency for each category. In this case there are three categories: limestone, granite and chert. The observed frequencies are 180, 186 and 234, and the expected frequencies are 200, 200 and 200. The value of χ^2 is therefore:

$$\chi^2 = \frac{(200-180)^2}{200} + \frac{(200-186)^2}{200} + \frac{(200-234)^2}{200}$$

$$= \frac{20^2}{200} + \frac{14^2}{200} + \frac{(-34)^2}{200} = \frac{400}{200} + \frac{196}{200} + \frac{1156}{200}$$

$$= 2.0 + 0.98 + 5.78 = 8.76$$

SIGNIFICANCE

A large value of χ^2 indicates that there is a large amount of difference between the observed and the expected frequencies, and would suggest that the null hypothesis can be rejected. It is, however, necessary before doing so to check whether the calculated value of χ^2 is indeed greater than the critical value given in the appropriate table (Appendix C5). It is necessary to know the degrees of freedom in order to use the tables, and these are given by the number of categories minus one. In this case there are three categories, limestone, granite and chert, and so the degrees of freedom are $3 - 1 = 2$.

Suppose a significance level of 0.01 is decided upon. From the appropriate row of Appendix C5 the critical value of χ^2 at the 0.01 significance level with 2 degrees of freedom is 9.21. This means that under the null hypothesis, that the sample has been taken from a population in which the three types of pebble occur in equal proportions, a sample outcome which gives rise to a value of χ^2 as large as, or larger than, 9.21 has a probability of occurrence of only 0.01. Since the calculated value of χ^2 is less than 9.21, the null hypothesis cannot be rejected at the 0.01 level.

The practical implications of this for the geomorphologist are that his probability of being wrong in rejecting the null hypothesis is something more than 0.01. In fact a further examination of the tables of χ^2 shows that the critical value at the 0.05 level with two degrees of freedom is 5.99. The calculated value of χ^2 is greater than this, and so the null hypothesis can be rejected at the 0.05 level.

In such a situation the researcher might well be tempted to choose the lower significance level (0.05), in order to be able to reject his null hypothesis. It is this sort of subjectivity which a good statistical geographer should try to avoid. The only safe rule is to choose a significance level **before** the test is carried out and then **stick to it**. Having said this, there is clearly considerable room for discussion about the choice of significance level in geographical situations.

The degrees of freedom for the one-sample chi square test can be thought of as the number of categories whose frequencies are free to vary or could be assigned arbitrarily. In the example the frequency for any two categories could be assigned arbitrarily. Once this had been done, however, the frequency of the third category would be fixed by the fact that the three frequencies must add up to 600. In any one-sample situation the number of frequencies which can be assigned arbitrarily is one less than the total number of categories.

RESTRICTIONS

Note that the data must be frequencies, i.e. the number of discrete objects occurring in different categories. The chi square test will normally give false results if applied to data concerning the proportions or percentages of occurrences in the categories. Also the categories must be mutually exclusive, so that one individual cannot possibly be counted in more than one category.

A further important restriction on this form of chi square test is that there should not be many categories for which the expected frequency is small. Quite what is meant in this context by 'many' and 'small' is a matter of dispute amongst statisticians, but two rules of thumb have been suggested (Siegel, 1956):

(1) if the number of categories is greater than 2, no more than 1/5 of the expected frequencies should be less than 5, and certainly none should be less than 1,

(2) if the number of categories is 2, both the expected frequencies should be 5 or larger.

If it is found when applying the test that the first rule has been broken it may be possible to combine categories until the offending expected frequencies have been suitably increased. In the pebble example, it would have been possible, for instance, to combine limestone and chert into a single category by adding together their observed and expected frequencies. It may be, however, that this process of combining cells could make nonsense of the underlying research

hypothesis. Certainly if the second rule is violated there is no alternative but to abandon the chi square test.

Exercise 4.6

The accompanying figure shows the distribution of electoral districts surrounding a city park, together with their population totals. A sample survey of 100 visitors to the park reveals that 13 come from electoral district A, 18 from B, 25 from C, 38 from D and 6 from E. It appears that visitors to the park are not drawn equally from the five surrounding E.Ds.

(a) Calculate the number of visitors (in a sample of 100) who would be expected to come from each E.D., if they had come in equal proportions in relation to the population of the E.D.

(b) Calculate the value of χ^2 for the difference between the observed and expected numbers. Is this difference significant at the 0.05 level?

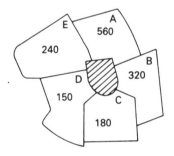

The Two-Sample Case

The chi square test can be used as a test of whether there is a difference between two samples of data expressed in frequency form. Table 4.11a shows a two-sample *contingency table*. One week's records for a driving test centre show that out of 100 women who took a first driving test during that week 75 passed and 25 failed. Of the 150 men who took a first test during the same period, 60 passed and 90 failed. If the men and women taking their tests during that week can be treated as a random cross-section of the population at large, then the men

Table 4.11 Chi Square Contingency Table for Driving Test Results

(a)		Result: Pass	Fail		(b)			
	Women	75	25	100		A	B	(A+B)
Samples:								
	Men	60	90	150		C	D	(C+D)
		135	115	250		(A+C)	(B+D)	N

and women can be treated as two random samples for the purpose of the chi square test.

Table 4.11a also gives the total number of passes and fails out of the 250 tests taken by the men and women together. The chi square test can be applied to this data to find out whether there is a significant difference between men and women in terms of their ability to pass a driving test at the first attempt.

The null hypothesis is that there is no difference between men and women in terms of their ability to pass a driving test at the first attempt. The alternative hypothesis is that there is a difference. From the sample data it can be seen that 3/4 of the women (75/100) passed at the first attempt, compared with only 2/5 of the men (60/150). The null hypothesis assumes this apparent difference is due to chance in the sampling process, and is not representative of the situation in the total population of first driving test candidates.

The test statistic for this situation, where there are only two samples and two categories, is calculated according to a special form of the χ^2 formula:

$$= \frac{n(|AD - BC| - n/2)^2}{(A+B)(C+D)(A+C)(B+D)} \quad (4.5)$$

where n is the total number of individuals in the two samples (the sum of all the frequencies), A, B, C and D refer to the frequencies in each of the cells of the contingency table 4.11b, and $|AD - BC|$ is the absolute value of the difference between A times D and B times C. As with the Student's t equation, the absolute value of the difference can be found by subtracting the smaller value from the larger. This equation is only applicable to a two by two contingency table. For the example data it gives:

$$\chi^2 = \frac{250(|75 \times 90 - 25 \times 60| - 250/2)^2}{(75+25)(60+90)(75+60)(25+90)}$$

$$= \frac{250(|6750 - 1500| - 125)^2}{100 \times 150 \times 135 \times 115}$$

$$= \frac{250(5250 - 125)^2}{232875000}$$

$$= \frac{250 \times 5125^2}{232875000} = \frac{250 \times 26265625}{232875000}$$

$$= \frac{6566406250}{232875000} = 28.197$$

The reader will not usually find himself working with such large numbers, but even if he does the calculations can be greatly simplified by the simple expedient of cancelling.

As with the one sample test, a large value of χ^2 tends to indicate a large difference and the same critical values apply (Appendix C5). The degrees of freedom for a two-sample test are given by the number of rows in the contingency table minus one multiplied by the number of columns minus one. In this case the degrees of freedom are therefore $(2-1) \times (2-1) = 1$.

If a significance level of 0.05 has been decided upon, the critical value of χ^2 is given in Appendix C5 as 3.84. Since the calculated value of χ^2 is considerably greater than this, the null hypothesis can be rejected at the 0.05 significance level.

In practical terms this means that, on the basis of the evidence of the chi square test, it is extremely unlikely that the observed difference in success rate between the samples of men and women drivers is due only to chance in the sampling process. It is far more likely that the sample data accurately reflect a 'real' difference between men and women drivers in the population at large.

Equation 4.5 is only appropriate for a two-sample chi square test in which there are only two categories, i.e. a two by two contingency table. If the two-sample chi square test is used when there are more than two categories, the more usual chi square formula is appropriate:

$$\chi^2 = \sum \frac{d^2}{e} \qquad (4.4)$$

This is the same equation as is used for the one-sample test. However, in a one-sample test the expected frequencies are known at the outset; in the two-sample case they must be calculated. How this is done can best be shown by example.

A survey of a random sample of 100 sixth formers is carried out to investigate the perception, or 'image', they have of their home town. As a result of a number of tests and exercises the students are put into one of three categories according to whether their images are mainly map-like, verbal or pictorial.

The researcher has a theory that the images held by students studying geography should be more map-like or pictorial than verbal. In order to test this theory the sample is split into geographers and others to give effectively two samples. The null hypothesis is that the two samples have come from a population in which there is no difference in image type between geographers and others. The observed difference is merely due to chance in the sampling process. The alternative hypothesis is that there is a real difference in the population between geographers and others. A significance level of 0.1 is decided upon.

The contingency table for this situation is shown in Table 4.12a. From the information given in this contingency table it can be said that 45/100 of the total set of students have 'map-like' images, 30/100 have 'verbal' images and 25/100 have 'pictorial' images. If there were no difference between the geographers and the others, these same proportions should be expected to apply to both samples. In other words 45/100 of the 28 geographers (i.e. about 13 of them) should have 'map-like' images.

Table 4.12 Chi Square Contingency Tables for perception Study

(a) *Observed Frequencies*

		Image type: Map-like	Verbal	Pictorial	
Samples:	Geographers	14	6	8	28
	Others	31	24	17	72
		45	30	25	100

(b) *Expected Frequencies*

	Map-like	Verbal	Pictorial	
Geographers	12.6 $\left(\dfrac{45 \times 28}{100}\right)$	8.4 $\left(\dfrac{30 \times 28}{100}\right)$	7.0 $\left(\dfrac{25 \times 28}{100}\right)$	28
Others	32.4 $\left(\dfrac{45 \times 72}{100}\right)$	21.6 $\left(\dfrac{30 \times 72}{100}\right)$	18.0 $\left(\dfrac{25 \times 72}{100}\right)$	72
	45	30	25	100

Similarly 45/100 of the 72 others (i.e. about 32 of them) should have 'map-like' images, and so on.

Table 4.12b shows all the expected frequencies, and in parentheses beneath each one the calculations required. Effectively the assumption is made that the proportion of individuals with 'map-like' images should be the same in each in the two samples as it is in the total set of 100 students, and the same applies to the proportions of individuals in the other categories. From Table 4.12b it should be clear that each expected frequency can be calculated by a simple mechanical rule:

$$\text{expected frequency} = \frac{\text{column total} \times \text{row total}}{\text{grand total}}$$

Having calculated the expected frequencies, the value of χ^2 can be calculated using Equation 4.4. In this case:

$$\chi^2 = \frac{(14-12.6)^2}{12.6} + \frac{(6-8.4)^2}{8.4} + \frac{(8-7.0)^2}{7.0}$$

$$+ \frac{(31-32.4)^2}{32.4} + \frac{(24-21.6)^2}{21.6}$$

$$+ \frac{(17-18.0)^2}{18.0}$$

$$= \frac{1.4^2}{12.6} + \frac{(-2.4)^2}{8.4} + \frac{1.0^2}{7.0} + \frac{(-1.4)^2}{32.4} + \frac{2.4^2}{21.6} + \frac{(-1.0)^2}{18.0}$$

$$= 0.156 + 0.686 + 0.143 + 0.060 + 0.267$$

$$+ 0.056 = 1.368$$

Degrees of freedom are calculated in the same way as in the previous example, as the number of rows minus one times the number of columns minus one. For this example the degrees of freedom are therefore $(2-1) \times (3-1) = 2$.

The critical value of χ^2 with 2 degrees of freedom at the 0.1 significance level is 4.60. In other words, a value of χ^2 as great as or greater than 4.60 could be obtained by chance under the null hypothesis with a probability of 0.1. Since the calculated value of χ^2 is less than this critical value, the null hypothesis must be accepted at the 0.1 probability level. According to the chi square test it is very improbable that the observed difference between the images held by the two samples of students is an indication of a 'real' difference between the two groups in the student population as a whole.

RESTRICTIONS

As in the one-sample cases, the chi square test can only validly be applied to a two-sample situation if the data are frequencies of occurrence of individuals in mutually exclusive categories. When Equation 4.4 is used, i.e. when the number of categories is greater than 2, dubious results can be obtained if there are many small expected frequencies. Once again a rule of thumb is suggested: not more than 1/5 of the expected frequencies should be less than 5, and none of the expected frequencies should be less than 1. If this rule is broken it may be possible to combine categories (i.e. columns of the contingency table) in a meaningful way and so increase the expected frequencies. If this process of amalgamation results in a reduction of the contingency table to 2 × 2, Equation 4.5 should be used.

Exercise 4.7

From an examination of residential preference data given in *Situations in Human Geography* (Cole, 1975, p. 40) it appears that coastal counties of England and Wales are perceived as being more desirable than inland counties. Of the 19 coastal counties 14 have preference scores of more than 30 and only 5 have preference scores of 30 or less. Of the 34 inland counties 15 have high preference scores and 19 have low scores. Use the chi square test to decide whether there is in fact a significant difference at the 0.05 level between coastal and inland counties in terms of their residential desirability.

Exercise 4.8

As part of a survey of shopping patterns a sample of 200 car owners are interviewed for comparison with an equivalent sample of 300 non car owners. In response to a question on frequency of shopping for groceries it is found that of the car owners 10 shop daily, 170 shop weekly and 20 shop monthly. The equivalent figures for the non car owners are 40, 180 and 80. Calculate the value of χ^2 for the difference between the two samples. Is this difference significant at the 0.01 level?

The Chi Square Test for Three or More Samples

The chi square test for three or more samples is a straightforward extension of the two-sample case. Suppose the perception study described above is extended to include two samples of adults, one of 'middle-aged' and one of 'old' people. The sixth formers are now taken as representing 'young' people. The results of this study are given in the contingency table in Table 4.13.

The null hypothesis is that there is no difference between the three age groups of people in the population at large. The alternative hypothesis is that the observed difference between the samples reflects a real difference in the population at large. A significance level of 0.05 is decided upon.

The expected frequencies are calculated in the same way as for the two-sample test, by multiplying together the appropriate row and column totals and dividing by the grand total. Table 4.13b gives the expected frequencies for all the cells of the contingency table.

The value of χ^2 can now be calculated using Equation 4.4:

$$\chi^2 = \frac{(45-33)^2}{33} + \frac{(30-26)^2}{26} + \frac{(25-41)^2}{41}$$

$$+ \frac{(11-16.5)^2}{16.5} + \frac{(8-13)^2}{13} + \frac{(31-20.5)^2}{20.5}$$

$$+ \frac{(10-16.5)^2}{16.5} + \frac{(14-13)^2}{13} + \frac{(26-20.5)^2}{20.5}$$

$$= \frac{144}{33} + \frac{16}{26} + \frac{256}{41} + \frac{30.25}{16.5} + \frac{25}{13} + \frac{110.25}{20.5}$$

$$+ \frac{42.25}{16.5} + \frac{1}{13} + \frac{30.25}{20.5}$$

$$= 4.36 + 0.62 + 6.24 + 1.83 + 1.92 + 5.38$$

$$+ 2.56 + 0.08 + 1.48$$

$$= 24.47$$

The degrees of freedom in this case are 4 (3 rows and 3 columns in the contingency table give $(3-1) \times (3-1)$ degrees of freedom). At the 0.05 significance level with 4 degrees of freedom the critical value of χ^2 is 9.49. Since the calculated value of χ^2 is greater than this, the null hypothesis can be rejected at the 0.05 significance level. The researcher can accept the alternative hypothesis that there is a real difference between age groups in terms of the types of images they possess.

Restrictions on the use of the chi square test with three or more samples are the same as those which apply in the two-sample case. The data must be in the form of frequencies of occurrence of individuals in mutually exclusive categories. Also not more than 1/5 of the expected frequencies should be less than 5, and none of the expected frequencies should be less than 1. Amalgamation of categories can again be used to overcome the problem of low expected frequencies.

4.6 The Kruskal–Wallis H Test

The Kruskal–Wallis H test is a test for deciding whether there is a significant difference between **three or more** samples. The H test has so far been little used in geography. It is, however, a useful alternative to analysis of variance (Section 4.7), since it is a **nonparametric** test and therefore does not rely on possibly unrealistic assumptions about the distribution of the variable. It can be applied to **ordinal** (ranked) data.

The null hypothesis of the H test is that the samples have been taken from populations with identical distributions. Any differences between the samples are due to chance variation inherent in the process of random sampling. The alternative hypothesis is that the samples have come from populations with different distributions, so that differences between the samples reflect real differences between the populations.

Table 4.14 contains data obtained from a sample of day visitors to three National Parks. The figures in columns a, c and e are the distance in kilometres travelled by each respondent on his journey to the park. The null hypothesis is that the three samples have come from identical populations. The alternative hypothesis is that

Table 4.13 Contingency Table for Extended Perception Test

(a) *Observed frequencies*

Image type:

		Map-like	Verbal	Pictorial	
	young	45	30	25	100
Samples:	middle-aged	11	8	31	50
	old	10	14	26	50
		66	52	82	200

(b) *Expected Frequencies*

33 $\left(\dfrac{66 \times 100}{200}\right)$	26 $\left(\dfrac{52 \times 100}{200}\right)$	41 $\left(\dfrac{82 \times 100}{200}\right)$	100
16.5 $\left(\dfrac{66 \times 50}{200}\right)$	13 $\left(\dfrac{52 \times 50}{200}\right)$	20.5 $\left(\dfrac{82 \times 50}{200}\right)$	50
16.5 $\left(\dfrac{66 \times 50}{200}\right)$	13 $\left(\dfrac{52 \times 50}{200}\right)$	20.5 $\left(\dfrac{82 \times 50}{200}\right)$	50
66	52	82	200

Table 4.14 Distances Travelled to National Parks by Sample Respondents

Park A		Park B		Park C	
a distance	b rank	c distance	d rank	e distance	f rank
23	7	11	3	110	18
15	4	17	5	84	16
42	12	31	11	85	17
8	1	27	9.5	45	13
10	2	63	14	64	15
18	6	27	9.5	25	8
$R_A = 32$ $n_A = 6$		$R_B = 52$ $n_B = 6$		$R_C = 87$ $n_C = 6$	

the three samples have come from different populations. A significance level of 0.05 is decided upon.

In order to apply the H test, the data must first be ranked, from lowest to highest. These rankings are given in columns b, d and f of Table 4.14. Note that they are the **overall** rankings of the sample values, not the rankings within the individual samples. As usual, identical values are given the mean of the ranking they would otherwise have received. The two values of 32 therefore become 9.5 and 9.5, instead of 9 and 10.

The sums of the ranks are then found for each sample. In this case these are the three values of R in Table 4.14. This information can now be used to calculate H from the following equation:

$$H = \frac{12}{N(N+1)} \sum \frac{R^2}{n} - 3(N+1) \qquad (4.6)$$

where N is the total number of individuals in all the samples, R is the sum of the ranks within a sample, and n is the number of individuals in that sample. The summation $\sum R^2/n$ means that the sum of all the values of R^2/n (one for each sample) must be found. For the data given in Table 4.14:

$$H = \frac{12}{18 \times 19} \left(\frac{32^2}{6} + \frac{52^2}{6} + \frac{87^2}{6} \right) - 3 \times 19$$

$$= 0.035(170.66 + 450.66 + 1261.5) - 57$$

$$= (0.035 \times 1882.82) - 57 = 8.90$$

When the samples all contain more than 5 individuals, the sampling distribution of H is virtually identical to the sampling distribution of χ^2. The appropriate degrees of freedom are given by the number of samples minus one; in this case 3 samples give 2 degrees of freedom.

At the 0.05 significance level the critical value of H can therefore be found in this case as the critical value of χ^2 with 2 degrees of freedom (Appendix C5), which is 5.99. Since the calculated value of H is larger than this critical value, the null hypothesis can be rejected. At the 0.05 significance level it is therefore safe to assume that the difference between the samples reflects a real difference between the populations of visitors to the three parks.

Correction for Tied Ranks

Strictly speaking, Equation 4.6 should be slightly modified if there are any tied ranks in the data. In the example there is only one set of tied ranks, the two values of 27 which are both ranked 9.5. To allow for tied ranks the calculated value of H is divided by a correction factor:

$$1 - \frac{\sum (t^3 - t)}{N^3 - N} \qquad (4.7)$$

where t is the number of individuals involved in each set of tied ranks, and N is the total number of individuals in all the samples. The summation $\sum (t^3 - t)$ means that the value of $(t^3 - t)$ must be calculated for each set of tied ranks, and the sum of all these values found.

For the example there is only one set of tied ranks, involving 2 values. The correction factor is therefore:

$$1 - \frac{(2^3 - 2)}{18^3 - 18} = 1 - \frac{6}{5814} = 0.999$$

The value of H from Equation 4.6 is 8.90; corrected for ties this becomes:

$$\frac{8.90}{0.999} = 8.91$$

Note that the effect of this correction for ties is quite small, and this is true even if quite a large number of tied ranks is involved. Also the effect of correcting for ties is to make the value of H larger and so increase the chance of rejecting the null hypothesis. In other words, if the correction factor is ignored the H test is erring on the side of caution. Unless more than a quarter of the values in the data set produce tied ranks, the effect of the correction factor is negligible (see Kruskal and Wallis, 1952, p. 587).

Small Samples

When small samples, containing fewer than 6 individuals are involved, the χ^2 approximation can no longer be used as a reliable means of testing the significance of H. In these situations the special sampling distribution of H must be used. Critical values of H for use with three small samples are given in Appendix C6.

The sampling distribution of H can be calculated directly, in a very similar way to the sampling distribution of the Mann–Whitney U statistic (Section 4.3). Table 4.15 shows all the 15 possible combinations of rankings in three samples containing 2, 2 and 1 individuals. The value of H produced by each combination of rankings is also shown. From this data it is possible to produce a frequency distribution of the values of H (Figure 4.6), and a cumulative probability distribution (Table 4.16). From this cumulative probability distribution it can be seen that, with sample sizes of 2, 2 and 1, a calculated value of H of as large as 3.0 or larger

is significant at the 0.333 probability level. In other words, it has a probability of 0.333 of occurring by chance under the null hypothesis.

It should also be clear from Table 4.16 that a large value of H is produced in a case where there is little overlap in the rankings between samples. The three combinations of rankings which produce the maximum value of H (3.6) are: 1/2, 3/4, 5; 1/2, 4/5, 3 and 2/3, 4/5, 1. In each of these sets of rankings there is no overlap at all between samples. This gives some insight into the way the H test 'works': by measuring the amount of overlap in rankings between samples.

Table 4.15 All Possible Combinations of Overall Ranks in Three Samples of 2, 2 and 1 Individuals in the Kruskal–Wallis H Test

Sample 1 (n = 2)	Sample 2 (n = 2)	Sample 3 (n = 1)	H
1, 2	3, 4	5	3.6
1, 2	3, 5	4	3.0
1, 2	4, 5	3	3.6
1, 3	2, 4	5	2.4
1, 3	2, 5	4	1.4
1, 3	4, 5	2	3.0
1, 4	2, 3	5	2.0
1, 4	2, 5	3	0.4
1, 4	3, 5	2	1.4
1, 5	2, 3	4	0.6
1, 5	2, 4	3	0.0
1, 5	3, 4	2	0.6
2, 3	4, 5	1	3.6
2, 4	3, 5	1	2.4
2, 5	3, 4	1	2.0

Table 4.16 Cumulative Probability Distribution of H

H	p
3.6	0.2000
3.0	0.3333
2.4	0.4667
2.0	0.6000
1.4	0.7333
0.6	0.8667
0.4	0.9333
0.0	1.0000

p gives the probability of getting a value of H as large as, or larger than, the one specified.

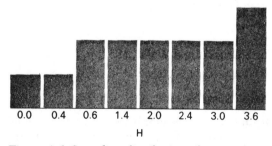

Figure 4.6 Sampling distribution of Kruskal–Wallis H

Exercise 4.9

As part of a biogeographical study soil samples are taken from three different areas of woodland. Laboratory analysis reveals that the water holding capacity of the samples is as follows:

Area A: 72, 58, 42, 85, 81, 64, 71, 68, 64, 73 (10 samples)

Area B: 36, 36, 30, 48, 43, 15, 44, 60 (8 samples)

Area C: 54, 57, 93, 64, 82, 34 (6 samples)

Use the Kruskal–Wallis H test to establish whether there is a significant difference in water holding capacity between the three areas at the 0.05 level.

4.7 Analysis of Variance

Analysis of variance (the F ratio test) is the standard **parametric** test of difference between **three or more samples**. Like other parametric tests, however, it is often used in situations where the rather rigid assumptions of such tests cannot be justified. Analysis of variance can only be applied to data measured on an **interval** scale.

The null hypothesis of analysis of variance is that the samples are taken from a common, normally distributed population (or from identical populations). The alternative hypothesis is that the samples come from populations with different distributions.

The rationale of analysis of variance is to find out whether there is more variation between the samples than within them. If the samples are taken at random from a common population (the null hypothesis), it is reasonable to expect the variation within the samples to be about the same as the variation between the samples, since both are reflections of the overall variation in the population. Any difference in these two measures of variation is merely due to chance in the sampling process. If the samples are taken from different populations (the alternative hypothesis), this is not a reasonable expecta-

tion, since the variation within each sample is a reflection of the variation within the particular population from which it has come. Variation between samples in this case is a reflection of the difference between the populations.

A social geographer is interested in differences in educational attainment in four areas of a city. As a preliminary exercise a random sample of primary school children is selected from each area, and each child is asked to complete a standard intelligence test. The IQ scores derived from these tests are listed in Table 4.17. Note that the samples are very small. In reality no one would put much faith in results obtained from such a small scale survey, but this simple example serves to demonstrate the application of analysis of variance.

The null hypothesis is that the sample IQ scores have come from a common, normally distributed population of IQ scores. Note that the assumption of a normal distribution is reasonable, since for any large number of people it is known that IQ scores are normally distributed. The alternative hypothesis is that the sample scores come from different, but still normally distributed, populations of IQ scores. A very cautious significance level of 0.001 is decided upon, since this area of research is rather controversial.

The first step in analysis of variance is to make

Table 4.17 IQ Scores for Sample Primary School Children

	Area A x	Area B x	Area C x	Area D x
	105	87	118	84
	102	93	95	85
	113	102	88	87
	96	98	101	96
	86	95	94	88
	97		92	94
	101			
	$\sum x = 700$	$\sum x = 475$	$\sum x = 588$	$\sum x = 534$
	$n = 7$	$n = 5$	$n = 6$	$n = 6$
	$\bar{x} = 100$	$\bar{x} = 95$	$\bar{x} = 98$	$\bar{x} = 89$

k (number of samples) = 4
N (total number of individuals) = 7 + 5 + 6 + 6 = 24
$\bar{x}_G$ (grand mean) = (700 + 475 + 588 + 534)/24 = 95.7

Table 4.18 Analysis of Variance Applied to IQ Scores of Sample Primary School Children

(*a*) *Calculation of Within Samples Variance Estimate*

Area A $(x-\bar{x})$	$(x-\bar{x})^2$	Area B $(x-\bar{x})$	$(x-\bar{x})^2$	Area C $(x-\bar{x})$	$(x-\bar{x})^2$	Area D $(x-\bar{x})$	$(x-\bar{x})^2$
5	25	−8	64	20	400	−5	25
2	4	−2	4	−3	9	−4	16
13	169	7	49	−10	100	−2	4
−4	16	3	9	3	9	7	49
−14	196	0	0	−4	16	−1	1
−3	9			−6	36	5	25
1	1						
$\overset{n}{\sum}(x-\bar{x})^2=420$		$\overset{n}{\sum}(x-\bar{x})^2=126$		$\overset{n}{\sum}(x-\bar{x})^2=570$		$\overset{n}{\sum}(x-\bar{x})^2=120$	

(Note that these are the deviations from the means for the data given in Table 4.17)

$$\hat{\sigma}_W^2=\frac{\overset{k}{\sum}\overset{n}{\sum}(x-\bar{x})^2}{N-k}=\frac{420+126+570+120}{24-4}=\frac{1236}{20}=61.8$$

(*b*) *Calculation of Between Samples Variance Estimate*

Area A $\bar{x}=100$ $n=7$ $n(\bar{x}-\bar{x}_G)^2=7(100-95.7)^2=7\times18.49=129.43$

Area B $\bar{x}=95$ $n=5$ $n(\bar{x}-\bar{x}_G)^2=5(95-95.7)^2=5\times0.49=2.45$

Area C $\bar{x}=98$ $n=6$ $n(\bar{x}-\bar{x}_G)^2=6(98-95.7)^2=6\times5.29=31.74$

Area D $\bar{x}=89$ $n=6$ $n(\bar{x}-\bar{x}_G)^2=6(89-95.7)^2=6\times44.89=269.34$

$$\hat{\sigma}_B^2=\frac{\overset{k}{\sum}n(\bar{x}-\bar{x}_G)^2}{k-1}=\frac{129.43+2.45+31.74+269.34}{4-1}=\frac{432.96}{3}=144.32$$

(*c*) *Analysis of Variance Table*

	Variance estimate	Degrees of freedom
Between samples	$\hat{\sigma}_B^2=144.32$	3 $(k-1)$
Within samples	$\hat{\sigma}_W^2=61.8$	20 $(N-k)$

$$F\text{ ratio}=\frac{\text{between samples variance estimate}}{\text{within samples variance estimate}}=\frac{144.32}{61.8}=2.34$$

Degrees of freedom for between samples variance estimate = 3

Degrees of freedom for within samples variance estimate = 20 two estimates of the variance of the hypothesized common population: the within samples variance estimate, and the between samples variance estimate.

The *within samples variance estimate* is calculated according to the following equation:

$$\hat{\sigma}_W^2 = \frac{\sum\limits^k \sum\limits^n (x - \bar{x})^2}{N - k} \qquad (4.8)$$

where $\hat{\sigma}_W^2$ is the within samples variance estimate, k is the number of samples, n is the number of individuals in each sample, N is the total number of individuals in all the samples put together, and $\bar{x}$ is the mean of each sample. The expression $\sum^n (x - \bar{x})^2$ means that for each sample the deviations from the mean are calculated, squared and summed. The summation sign $\sum^k$ means that all the sums of squares, one for each sample, are added together. Table 4.18a shows the calculation of $\hat{\sigma}_W^2$ for the example data, giving a value of 61.8.

The *between samples variance estimate* can now be calculated as:

$$\hat{\sigma}_B^2 = \frac{\sum\limits^k n(\bar{x} - \bar{x}_G)^2}{k - 1} \qquad (4.9)$$

where $\hat{\sigma}_B^2$ is the between samples variance estimate, n is the number of individuals in a sample, k is the number of samples, $\bar{x}$ is the mean of a sample, and $\bar{x}_G$ is the grand mean of all the data values. The expression $n(\bar{x} - \bar{x}_G)^2$ means that for each sample the deviation of the sample mean from the grand mean is found, squared and then multiplied by the number of individuals in the sample. The summation $\sum^k$ means that these values are added together (there will be one for each sample).

Table 4.18b shows the calculation of $\hat{\sigma}_B^2$ for the example data, giving a value of 144.32. What happens in the calculation of the between samples variance estimate is that the variation within each sample is eliminated in order to measure the variation between the samples. This is done by effectively replacing each sample value by the mean of the sample, and measuring the deviation of the sample mean from the grand mean. In Table 4.18b the calculation of $n(x - \bar{x}_G)^2$ for area A, for example,

involves finding the square of the deviation of the mean of that sample (100) from the grand mean (95.7) and multiplying by the number of individuals in the sample. This is equivalent to replacing each sample value by the sample mean.

Having calculated two estimates of the population variance, the question now is: how probable is it that these two values are estimates of the same population variance? In order to answer this question a statistic known as the F ratio is calculated:

$$F = \frac{\text{between samples variance estimate}}{\text{within samples variance estimate}} \qquad (4.10)$$

For the example data:

$$F = \frac{\hat{\sigma}_B^2}{\hat{\sigma}_W^2} = \frac{144.32}{61.8} = 2.34$$

Significance

Tables of critical values of F are available (Appendix C7), and these can be consulted in order to find out whether a value of 2.34 is significant at the 0.001 level. In order to use these tables two sets of degrees of freedom must be known: the degrees of freedom for the two variance estimates.

The degrees of freedom for analysis of variance are calculated as follows:

(1) for the between samples variance estimate ($\hat{\sigma}_B^2$) they are the number of samples minus one (k − 1).

(2) for the within samples variance estimate ($\hat{\sigma}_W^2$) they are the total number of individuals in the data minus the number of samples (N − k)

Since the calculations involved in analysis of variance are quite complex, it is advisable to draw up an analysis of variance table summarizing the values obtained. For the example data this is done in Table 4.18c.

The critical value of F at the 0.001 significance level with 3 and 20 degrees of freedom can now be found by consulting the appropriate section of Appendix C7. Note that there is one section for each significance level. The critical value in this case is 8.10. Since the calculated value of F is less than this, the null hypothesis must be accepted at the 0.001 probability level.

The researcher must conclude that the sample IQ scores have in fact all come from a common population (or identical populations). On the basis of the F test there is no support for the contention that the differences in IQ between the four sample sets of shoolchildren are representative of real differences in the primary school populations of those areas.

4.8 Computer Programs

Program 3 — Kolmogorov–Smirnov Test

The program performs a Kolmogorov–Smirnov test between an observed **frequency** distribution and a theoretical **probability** distribution.

Program Listing

```
10 REM - KOLMOGOROV-SMIRNOV TEST
20 PRINT"---------------------"
30 PRINT"KOLMOGOROV-SMIRNOV TEST"
40 PRINT"---------------------"
50 PRINT
60 INPUT"Number of classes",NX
70 PRINT
80 N=NX+NX
90 DIM X(N)
100 PRINT
110 PRINT"Now type in observed FREQUENCIES:"
120 PRINT"(Mistakes can be corrected later)"
130 FOR I=1 TO NX
140 PRINT"Class ";I;" ";
150 INPUT X(I)
160 NEXT I
170 PRINT
180 PRINT"Now type in theoretical PROBABILITIES:"
190 PRINT"(Mistakes can be corrected later)"
200 FOR I=NX+1 TO N
210 PRINT"Class ";I-NX;" ";
220 INPUT X(I)
230 NEXT I
240 N1=1
250 N2=NX
260 V$="observed frequency"
270 GOSUB 500
280 N1=NX+1
290 N2=N
300 V$="theoretical probability"
310 GOSUB 500
320 T=0
330 FOR I=1 TO NX
340 T=T+X(I)
350 NEXT I
351 PRINT
352 PRINT"---------------------"
```

```
353 PRINT"KOLMOGOROV-SMIRNOV TEST"
354 PRINT"---------------------"
355 PRINT
356 PRINT"CUMULATIVE PROBABILITIES:"
357 PRINT
358 PRINT"   Observed   Theoretical   Difference"
359 PRINT
360 P1=0
370 P2=0
380 DM=0
390 FOR I=1 TO NX
400 P1=P1+X(I)/T
410 P2=P2+X(I+NX)
420 D=ABS(P1-P2)
430 IF D>DM DM=D
440 PRINT P1,P2,D
450 NEXT I
460 PRINT
470 PRINT"D = ";DM
480 PRINT"with ";T;" degrees of freedom"
490 STOP
500 REM - DATA CHECKING ROUTINE
510 PRINT
520 PRINT"Do you want to check/correct ";V$;
530 INPUT" values (Y/N)";Q$
540 IF Q$="N" GOTO 650
550 IF Q$<>"Y" GOTO 520
560 K=0
570 PRINT"Press RETURN if value is correct";
580 PRINT" - otherwise type correct value"
590 FOR I=N1 TO N2
600 K=K+1
610 PRINT K;" ";X(I);
620 INPUT A$
630 IF A$<>"" X(I)=VAL(A$)
640 NEXT I
650 RETURN
```

Exercise 4.10

Calculate the value of F for the difference in soil water holding capacity between three woodland areas (data given in Exercise 4.9). Is the difference significant at the 0.05 level?

Example Output

```
------------------------
KOLMOGOROV-SMIRNOV TEST
------------------------

Number of classes?11


Now type in observed FREQUENCIES:
(Mistakes can be corrected later)
Class 1 ?30
Class 2 ?25
Class 3 ?15
Class 4 ?12
Class 5 ?5
Class 6 ?6
Class 7 ?2
Class 8 ?1
Class 9 ?0
Class 10 ?0
Class 11 ?4

Now type in theoretical PROBABILITIES:
(Mistakes can be corrected later)
Class 1 ?0.13
Class 2 ?0.26
Class 3 ?0.27
Class 4 ?0.19
Class 5 ?0.10
Class 6 ?0.04
Class 7 ?0.01
Class 8 ?0.00
Class 9 ?0.00
Class 10 ?0.00
Class 11 ?0.00

Do you want to check/correct observed frequency values (Y/N)?N

Do you want to check/correct theoretical probability values (Y/N)?N


------------------------
KOLMOGOROV-SMIRNOV TEST
------------------------

CUMULATIVE PROBABILITIES:

   Observed  Theoretical  Difference

     0.30       0.13        0.17
     0.55       0.39        0.16
     0.70       0.66        0.04
     0.82       0.85        0.03
     0.87       0.95        0.08
     0.93       0.99        0.06
     0.95       1.00        0.05
     0.96       1.00        0.04
     0.96       1.00        0.04
     0.96       1.00        0.04
     1.00       1.00        0.00

D = 0.17
with 100 degrees of freedom

STOP at line 490
```

The data used are from Table 4.1.

Program 4 — Mann–Whitney U Test

The program performs a Mann–Whitney U test
of the difference between two samples.

Program Listing

```
10 REM - MANN-WHITNEY U TEST
20 PRINT"-------------------"
30 PRINT"MANN-WHITNEY U TEST"
40 PRINT"-------------------"
50 PRINT
60 INPUT"Number of values of X",NX
70 PRINT
80 INPUT"Number of values of Y",NY
90 N=NX+NY
100 DIM X(N),N(N),R(N)
110 PRINT
120 PRINT"Now type in values of X:"
130 PRINT"(mistakes can be corrected later)"
140 FOR I=1 TO NX
150 PRINTI;" ";
160 INPUT X(I)
170 N(I)=I
180 NEXT I
190 PRINT
200 PRINT"Now type in values of Y:"
210 PRINT"(mistakes can be corrected later)"
220 FOR I=NX+1 TO NX+NY
230 PRINT I-NX;" ";
240 INPUT X(I)
250 N(I)=I
260 NEXT I
270 N1=1
280 N2=NX
290 V$="X"
300 GOSUB 1000
310 N1=NX+1
320 N2=N
330 V$="Y"
340 GOSUB 1000
350 GOSUB 830
360 R(N(N))=N
370 NT=0
380 FOR I=1 TO N-1
390 FOR J=I+1 TO N
400 IF X(I)<>X(J) GOTO 430
410 NT=NT+1
420 NEXT J
430 IF NT<>0 GOTO 470
440 R(N(I))=I
450 R(N(J))=J
460 GOTO 520
470 FOR K=I TO J-1
480 R(N(K))=(I+J-1)/2
490 NEXT K
500 I=I+NT
510 NT=0
520 NEXT I
530 FOR I=1 TO NX
540 RX=RX+R(I)
550 NEXT I
560 FOR I=NX+1 TO N
570 RY=RY+R(I)
580 NEXT I
```

```
590 PRINT
600 PRINT"-------------------"
610 PRINT"MANN-WHITNEY U TEST"
620 PRINT"-------------------"
630 PRINT
640 PRINT"Mean rank of X = ";RX/NX
650 PRINT"Mean rank of Y = ";RY/NY
660 PRINT
670 IF RX/NX<RY/NY PRINT" (i.e. X<Y)"
680 IF RX/NX>RY/NY PRINT" (i.e. X>Y)"
690 IF RX/NX=RY/NY PRINT" (i.e. X=Y)"
700 UX=NX*NY+NX*(NX+1)/2-RX
710 UY=NX*NY+NY*(NY+1)/2-RY
720 UM=UX
730 IF UY<UM UM=UY
740 PRINT
750 PRINT"Umin (two-tailed test - H1: X<>Y) = ";UM
760 PRINT
770 PRINT"Ux   (one-tailed test - H1: X>Y) = ";UX
780 PRINT
790 PRINT"Uy   (one-tailed test - H1: X<Y) = ";UY
800 PRINT
810 PRINT"with ";NX;" and ";NY;" degrees of freedom"
820 STOP
830 REM - BUBBLE SORT RANKING ROUTINE
840 Z=0
850 FOR I=1 TO N-1
860 J=I+1
870 IF X(I)<X(J) GOTO 970
880 IF X(I)>X(J) GOTO 900
890 GOTO 970
900 D=X(J)
910 DN=N(J)
920 N(J)=N(I)
930 X(J)=X(I)
940 N(I)=DN
950 X(I)=D
960 Z=1
970 NEXT I
980 IF Z=1 GOTO 840
990 RETURN
1000 REM - DATA CHECKING ROUTINE
1010 PRINT
1020 PRINT"Do you want to check/correct ";V$;
1030 INPUT" values (Y/N)",Q$
1040 IF Q$="N" GOTO 1150
1050 IF Q$<>"Y" GOTO 1020
1060 K=0
1070 PRINT"Press RETURN if value is correct";
1080 PRINT" - otherwise type correct value"
1090 FOR I=N1 TO N2
1100 K=K+1
1110 PRINT K;" ";X(I);
1120 INPUT A$
1130 IF A$<>"" X(I)=VAL(A$)
1140 NEXT I
1150 RETURN
```

Example Output

```
--------------------
MANN-WHITNEY U TEST
--------------------

Number of values of X?6

Number of values of Y?5

Now type in values of X:
(mistakes can be corrected later)
        1 ?15.2
        2 ?10.7
        3 ?8.6
        4 ?9.3
        5 ?12.4
        6 ?11.1

Now type in values of Y:
(mistakes can be corrected later)
        1 ?8.4
        2 ?9.3
        3 ?8.7
        4 ?10.2
        5 ?10.0

Do you want to check/correct X values (Y/N)?N

Do you want to check/correct Y values (Y/N)?N

--------------------
MANN-WHITNEY U TEST
--------------------

Mean rank of X = 7.41666667
Mean rank of Y = 4.3

 (i.e. X>Y)

Umin (two-tailed test - H1: X<>Y) = 6.5

Ux   (one-tailed test - H1: X>Y) = 6.5

Uy   (one-tailed test - H1: X<Y) = 23.5

with 6 and 5 degrees of freedom

STOP at line 820
```

The data used are from the worked example in
the text (Table 4.9).

Program 5 — Student's t Test

The program performs a Student's t test of the difference between two samples.

Program Listing

```
10 REM - STUDENT'S T TEST
20 PRINT"----------------"
30 PRINT"STUDENT'S T TEST"
40 PRINT"----------------"
50 PRINT
60 INPUT"Number of values of X",NX
70 PRINT
80 INPUT"Number of values of Y",NY
90 N=NX+NY
100 XM=0
110 YM=0
120 XX=0
130 YY=0
140 DIM X(N)
150 PRINT
160 PRINT"Now type in values of X:"
170 PRINT"(Mistakes can be corrected later)"
180 FOR I=1 TO NX
190 PRINT I;" ";
200 INPUT X(I)
210 NEXT I
220 PRINT
230 PRINT"Now type in values of Y:"
240 PRINT"(Mistakes can be corrected later)"
250 FOR I=NX+1 TO N
260 PRINT I-NX;" ";
270 INPUT X(I)
280 NEXT I
290 N1=1
300 N2=NX
310 V$="X"
320 GOSUB 650
330 FOR I=1 TO NX
340 XM=XM+X(I)
350 XX=XX+X(I)^2
360 NEXT I
370 XM=XM/NX
380 SX=(XX/NX-XM^2)/(NX-1)
390 N1=NX+1
400 N2=N
410 V$="Y"
420 GOSUB 650
430 FOR I=NX+1 TO N
440 YM=YM+X(I)
450 YY=YY+X(I)^2
460 NEXT I
470 YM=YM/NY
480 SY=(YY/NY-YM^2)/(NY-1)
490 T=ABS(XM-YM)/SQR(SX+SY)
500 PRINT
510 PRINT"----------------"
520 PRINT"STUDENT'S T TEST"
530 PRINT"----------------"
540 PRINT
550 PRINT"Mean of X = ";XM
560 PRINT"Standard deviation of X = ";SQR(XX/NX-XM^2)
570 PRINT
580 PRINT"Mean of Y = ";YM
590 PRINT"Standard deviation of Y = ";SQR(YY/NY-YM^2)
600 PRINT
610 PRINT"T = ";T
620 DF=INT((SX+SY)^2/(SX^2/(NX-1)+SY^2/(NY-1)))
630 PRINT"with ";DF;" degrees of freedom"
640 STOP
650 REM - DATA CHECKING ROUTINE
660 PRINT
670 PRINT"Do you want to check/correct ";V$;
680 INPUT" values (Y/N)";Q$
690 IF Q$="N" GOTO 800
700 IF Q$<>"Y" GOTO 670
710 K=0
720 PRINT"Press RETURN if value is correct";
730 PRINT" - otherwise type correct value"
740 FOR I=N1 TO N2
750 K=K+1
760 PRINT K;" ";X(I);
770 INPUT A$
780 IF A$<>"" X(I)=VAL(A$)
790 NEXT I
800 RETURN
```

Example Output

```
------- ----------
STUDENT'S T TEST
----------------

Number of values of X?10

Number of values of Y?10

Now type in values of X:
(Mistakes can be corrected later)
        1 ?37
        2 ?46
        3 ?53
        4 ?50
        5 ?48
        6 ?27
        7 ?65
        8 ?39
        9 ?28
       10 ?42

Now type in values of Y:
(Mistakes can be corrected later)
        1 ?48
        2 ?63
        3 ?62
        4 ?51
        5 ?46
        6 ?43
        7 ?39
        8 ?21
        9 ?57
       10 ?60

Do you want to check/correct X values (Y/N)?N

Do you want to check/correct Y values (Y/N)?N

----------------
STUDENT'S T TEST
----------------

Mean of X = 43.5
Standard deviation of X = 10.9476025

Mean of Y = 49
Standard deviation of Y = 12.1819539

T = 1.00742814
with 17 degrees of freedom

STOP at line 640
```

The data used are from the worked example in
the text (Table 4.9).

Program 6 — Chi Square Test

This program performs a chi square test for two or more samples. For a two-sample by two category test the special formula for chi square (Equation 4.5) is automatically used.

Program Listing

```
10 REM - CHI SQUARE TEST
20 PRINT"---------------"
30 PRINT"CHI SQUARE TEST"
40 PRINT"---------------"
50 PRINT
60 INPUT"Number of samples",S
70 PRINT
80 INPUT"Number of categories in each sample",NS
90 DIM X(S,NS),CT(NS),RT(S)
100 FOR I=1 TO S
110 PRINT
120 PRINT"Now type in values for sample ";I;":"
130 PRINT"(mistakes can be corrected later)"
140 K=1
150 FOR J=1 TO NS
160 PRINT"Category ";J;" ";
170 INPUT X(I,J)
180 NEXT J
190 RT(I)=0
200 NEXT I
210 FOR K=1 TO S
220 N1=1
230 N2=NS
240 GOSUB 770
250 CT(K)=0
260 NEXT K
270 PRINT
280 PRINT"---------------"
290 PRINT"CHI SQUARE TEST"
300 PRINT"---------------"
310 PRINT
320 IF S<>2 OR NS<>2 GOTO 450
330 AD=X(1,1)*X(2,2)
340 BC=X(1,2)*X(2,1)
350 AB=X(1,1)+X(1,2)
360 CD=X(2,1)+X(2,2)
370 AC=X(1,1)+X(2,1)
380 BD=X(1,2)+X(2,2)
390 N=AB+CD
400 CH=N*(ABS(AD-BC)-N/2)^2
410 CH=CH/(AB*CD*AC*BD)
420 PRINT"Chi square = ";CH
430 PRINT"with 1 degree of freedom"
440 STOP
450 GT=0
460 FOR I=1 TO S
470 FOR J=1 TO NS
480 RT(I)=RT(I)+X(I,J)
490 CT(J)=CT(J)+X(I,J)
500 GT=GT+X(I,J)
510 NEXT J
520 NEXT I
530 PRINT"EXPECTED VALUES"
540 NE=0
550 NN=0
560 FOR I=1 TO S
570 PRINT
580 PRINT"Sample ";I;":"
590 FOR J=1 TO NS
600 E=CT(J)*RT(I)/GT
610 CH=CH+(X(I,J)-E)^2/E
620 PRINT"Category ";J;"   ";E
630 IF E<5 NE=NE+1
640 IF E<1 NN=NN+1
650 NEXT J
660 NEXT I
670 E=NE*100/(NS*S)
680 PRINT
690 PRINT"% of expected values < 5 = ";E
700 E=NN*100/(NS*S)
710 PRINT"% of expected values < 1 = ";E
720 DF=(S-1)*(NS-1)
730 PRINT
740 PRINT"Chi square = ";CH
750 PRINT"with ";DF;" degrees of freedom"
760 STOP
770 REM - DATA CHECKING ROUTINE
780 PRINT
790 PRINT"Do you want to check/correct sample ";K;
800 INPUT" values (Y/N)";Q$
810 IF Q$="N" GOTO 900
820 IF Q$<>"Y" GOTO 790
830 PRINT"Press RETURN if value is correct";
840 PRINT" - otherwise type correct value"
850 FOR I=N1 TO N2
860 PRINT I;" ";X(K,I);
870 INPUT A$
880 IF A$<>"" X(K,I)=VAL(A$)
890 NEXT I
900 RETURN
```

Example Output 1

```
----------------
CHI SQUARE TEST
----------------

Number of samples?2

Number of categories in each sample?2

Now type in values for sample 1:
(mistakes can be corrected later)
Category 1 ?75
Category 2 ?25

Now type in values for sample 2:
(mistakes can be corrected later)
Category 1 ?60
Category 2 ?90

Do you want to check/correct sample 1 values (Y/N)?N

Do you want to check/correct sample 2 values (Y/N)?N

----------------
CHI SQUARE TEST
----------------

Chi square = 28.1971283
with 1 degree of freedom

STOP at line 440
```

The data used are from Table 4.11. This is a two sample by two category test.

Example Output 2

```
----------------
CHI SQUARE TEST
----------------

Number of samples?2

Number of categories in each sample?3

Now type in values for sample 1:
(mistakes can be corrected later)
Category 1 ?14
Category 2 ?6
Category 3 ?8

Now type in values for sample 2:
(mistakes can be corrected later)
Category 1 ?31
Category 2 ?24
Category 3 ?17

Do you want to check/correct sample 1 values (Y/N)?N

Do you want to check/correct sample 2 values (Y/N)?N

----------------
CHI SQUARE TEST
----------------

EXPECTED VALUES

Sample 1:
Category 1  12.6
Category 2  8.4
Category 3  7

Sample 2:
Category 1  32.4
Category 2  21.6
Category 3  18

% of expected values < 5 = 0
% of expected values < 1 = 0

Chi square = 1.36684303
with 2 degrees of freedom

STOP at line 760
```

The data used are from the perception study example (Table 4.12), which involves a two by three contingency table. In this case the program uses the normal form of chi square (Equation 4.4). The expected values are output, and the program calculates the percentage of expected values that are less than five and the percentage less than one.

Program 7 — Kruskal–Wallis H Test

This program performs a Kruskal–Wallis H test for two or more samples. The correction for tied ranks is applied when necessary.

Program Listing

```
10 REM - KRUSKAL-WALLIS H TEST
20 PRINT"--------------------"
30 PRINT"KRUSKAL-WALLIS H TEST"
40 PRINT"--------------------"
50 PRINT
60 INPUT"Number of samples",S
70 DIM NT(S),NS(S)
80 NT(0)=0
90 FOR I=1 TO S
100 PRINT
110 PRINT"Number of values in sample ";I;
120 INPUT NS(I)
130 NT(I)=NT(I-1)+NS(I)
140 NEXT I
150 N=NT(S)
160 DIM X(N),R(N),N(N)
170 FOR I=1 TO S
180 PRINT
190 PRINT"Now type in values for sample ";I;":"
200 PRINT"(mistakes can be corrected later)"
210 K=1
220 FOR J=NT(I-1)+1 TO NT(I)
230 PRINT K;" ";
240 INPUT X(J)
250 K=K+1
260 N(J)=J
270 NEXT J
280 NEXT I
290 FOR K=1 TO S
300 N1=NT(K-1)+1
310 N2=NT(K)
320 GOSUB 960
330 NEXT K
340 GOSUB 790
350 R(N(N))=N
360 RT=0
370 T=0
380 FOR I=1 TO N-1
390 FOR J=I+1 TO N
400 IF X(I)<>X(J) GOTO 430
410 RT=RT+1
420 NEXT J
430 IF RT<>0 GOTO 470
440 R(N(I))=I
450 R(N(J))=J
460 GOTO 530
470 FOR K=I TO J-1
480 R(N(K))=(I+J-1)/2
490 NEXT K
500 T=T+(RT+1)^3-RT-1
510 I=I+RT
520 RT=0
530 NEXTI
540 R=0
550 FOR I=1 TO S
560 R2=0
```

```
570 FOR J=NT(I-1)+1 TO NT(I)
580 R2=R2+R(J)
590 NEXT J
600 R=R+R2^2/NS(I)
610 R(I)=R2/NS(I)
620 NEXT I
630 PRINT
640 PRINT"--------------------"
650 PRINT"KRUSKAL-WALLIS H TEST"
660 PRINT"--------------------"
670 PRINT
680 PRINT"Mean ranks of samples:"
690 FOR I=1 TO S
700 PRINT"Sample ";I;": ";R(I)
710 NEXT I
720 H=12/(N*(N+1))*R-3*(N+1)
730 H=H/(1-T/(N^3-N))
740 PRINT
750 PRINT"H (Chi square) = ";H
760 PRINT"(corrected for ties)"
770 PRINT"Degrees of freedom for Chi square = ";S-1
780 STOP
790 REM - BUBBLE SORT RANKING ROUTINE
800 Z=0
810 FOR I=1 TO N-1
820 J=I+1
830 IF X(I)<X(J) GOTO 930
840 IF X(I)>X(J) GOTO 860
850 GOTO 930
860 D=X(J)
870 DN=N(J)
880 N(J)=N(I)
890 X(J)=X(I)
900 N(I)=DN
910 X(I)=D
920 Z=1
930 NEXT I
940 IF Z=1 GOTO 800
950 RETURN
960 REM - DATA CHECKING ROUTINE
970 PRINT
980 PRINT"Do you want to check/correct sample ";K;
990 INPUT" values (Y/N)",Q$
1000 IF Q$="N" GOTO 1110
1010 IF Q$<>"Y" GOTO 980
1020 J=0
1030 PRINT"Press RETURN if value is correct";
1040 PRINT" - otherwise type correct value"
1050 FOR I=N1 TO N2
1060 J=J+1
1070 PRINT J;" ";X(I);
1080 INPUT A$
1090 IF A$<>"" X(I)=VAL(A$)
1100 NEXT I
1110 RETURN
```

Example Output

```
----------------------
KRUSKAL- WALLIS H TEST
----------------------

Number of samples?3

Number of values in sample 1?6

Number of values in sample 2?6

Number of values in sample 3?6

Now type in values for sample 1:
(mistakes can be corrected later)
        1 ?23
        2 ?15
        3 ?42
        4 ?8
        5 ?10
        6 ?18

Now type in values for sample 2:
(mistakes can be corrected later)
        1 ?11
        2 ?17
        3 ?31
        4 ?27
        5 ?63
        6 ?27

Now type in values for sample 3:
(mistakes can be corrected later)
        1 ?110
        2 ?84
        3 ?85
        4 ?45
        5 ?64
        6 ?25

Do you want to check/correct sample 1 values (Y/N)?N

Do you want to check/correct sample 2 values (Y/N)?N

Do you want to check/correct sample 3 values (Y/N)?N


----------------------
KRUSKAL-WALLIS H TEST
----------------------

Mean ranks of samples:
Sample 1: 5.33333333
Sample 2: 8.66666667
Sample 3: 14.5

H (Chi square) = 9.07369145
(corrected for ties)
Degrees of freedom for Chi square = 2

STOP at line 780
```

The data used are from the worked example in
the text (Table 4.14).

Program 8 — Analysis of Variance

This program performs an analysis of variance
(F test) for two or more samples.

Program Listing

```
10 REM - ANALYSIS OF VARIANCE (F TEST)
20 PRINT"--------------------------"
30 PRINT"ANALYSIS OF VARIANCE (F TEST)"
40 PRINT"--------------------------"
50 PRINT
60 INPUT"Number of samples",S
70 DIM NS(S),NT(S)
80 NT(0)=0
90 FOR I=1 TO S
100 PRINT
110 PRINT"Number of values in sample ";I;
120 INPUT NS(I)
130 NT(I)=NT(I-1)+NS(I)
140 NEXT I
150 N=NT(S)
160 DIM X(N)
170 FOR I=1 TO S
180 PRINT
190 PRINT"Now type in values for sample ";I;":"
200 PRINT"(mistakes can be corrected later)"
210 K=1
220 FOR J=NT(I-1)+1 TO NT(I)
230 PRINT K;" ";
240 INPUT X(J)
250 K=K+1
260 NEXT J
270 NEXT I
280 FOR K=1 TO S
290 N1=NT(K-1)+1
300 N2=NT(K)
310 GOSUB 660
320 NEXT K
330 PRINT
340 PRINT"--------------------------"
350 PRINT"ANALYSIS OF VARIANCE (F TEST)"
360 PRINT"--------------------------"
370 FOR I=1 TO S
380 XS=0
390 XX=0
400 FOR J=NT(I-1)+1 TO NT(I)
410 XS=XS+X(J)
420 XX=XX+X(J)^2
430 NEXT J
440 SD=SQR(XX/NS(I)-(XS/NS(I))^2)
450 PRINT
460 PRINT"Sample ";I;":"
470 PRINT"Mean = ";XS/NS(I)
480 PRINT"Standard deviation = ";SD
490 GT=GT+XS
500 G2=G2+XX
510 SB=SB+XS^2/NS(I)
520 NEXT I
530 ST=G2-GT^2/N
540 SB=SB-GT^2/N
550 SW=ST-SB
560 VB=SB/(S-1)
570 VW=SW/(N-S)
580 PRINT
590 PRINT"Between samples variance = ";VB
600 PRINT"Within samples variance = ";VW
610 F=VB/VW
620 PRINT
630 PRINT"F = ";F
640 PRINT"with ";(S-1);" and ";(N-S);" degrees of freedom"
650 STOP
660 REM - DATA CHECKING ROUTINE
670 PRINT
680 PRINT"Do you want to check/correct sample ";K;
690 INPUT" values (Y/N)";Q$
700 IF Q$="N" GOTO 810
710 IF Q$<>"Y" GOTO 680
720 J=0
730 PRINT"Press RETURN if value is correct";
740 PRINT" - otherwise type correct value"
750 FOR I=N1 TO N2
760 J=J+1
770 PRINT J;" ";X(I);
780 INPUT A$
790 IF A$<>"" X(I)=VAL(A$)
800 NEXT I
810 RETURN
```

Example Output

```
------------------------------
ANALYSIS OF VARIANCE (F TEST)
------------------------------

Number of samples?4

Number of values in sample 1?7

Number of values in sample 2?5

Number of values in sample 3?6

Number of values in sample 4?6

Now type in values for sample 1:
(mistakes can be corrected later)
        1 ?105
        2 ?102
        3 ?113
        4 ?96
        5 ?86
        6 ?97
        7 ?101

Now type in values for sample 2:
(mistakes can be corrected later)
        1 ?87
        2 ?93
        3 ?102
        4 ?98
        5 ?95

Now type in values for sample 3:
(mistakes can be corrected later)
        1 ?118
        2 ?95
        3 ?88
        4 ?101
        5 ?94
        6 ?92

Now type in values for sample 4:
(mistakes can be corrected later)
        1 ?84
        2 ?85
        3 ?87
        4 ?96
        5 ?88
        6 ?94

Do you want to check/correct sample 1 values (Y/N)?N

Do you want to check/correct sample 2 values (Y/N)?N

Do you want to check/correct sample 3 values (Y/N)?N

Do you want to check/correct sample 4 values (Y/N)?N
```

```
------------------------------
ANALYSIS OF VARIANCE (F TEST)
------------------------------

Sample 1:
Mean = 100
Standard deviation = 7.74596669

Sample 2:
Mean = 95
Standard deviation = 5.01996023

Sample 3:
Mean = 98
Standard deviation = 9.74679435

Sample 4:
Mean = 89
Standard deviation = 4.47213596

Between samples variance = 144.319438
Within samples variance = 61.8

F = 2.33526598
with 3 and 20 degrees of freedom

STOP at line 650
```

The data used are from the worked example in the text (Table 4.17).

5 Relationships

Chapter 4 was concerned with techniques for measuring the degree of difference between two or more samples based on measured values of one variable. In Section 4.4, for example, Student's t test was used to see if there was a significant difference between two housing estates (the samples) in terms of the ages of heads of household (the variable). An equally useful application of statistics is in describing the relationship between two variables measured for a common sample of individuals.

In an area of heathland a biogeographer notices that there are large numbers of alder trees and saplings where the ground is rather marshy, whilst alder trees are virtually absent from drier parts of the heath. There appears to be a relationship between the occurrence of alder trees and the wetness of the ground. In order to test this hypothesis a random sample of 25 quadrats is selected throughout the heath. At each one the number of alder trees and saplings is counted and the moisture content of the soil measured. So for each of the individuals (quadrats) in the sample two variables are measured, number of alders and soil moisture content.

These data can be expressed as scatter plots, and a number of possible sets of sample results are shown in this way in Figure 5.1. Figure 5.1a illustrates a situation in which there is a direct relationship between the two variables. Quadrats with high values of moisture content tend to contain large numbers of alders, whereas quadrats with low moisture content contain few alders. The situation shown in Figure 5.1b is one of an inverse relationship. High values of mois-ture content are associated with low numbers of alders and vice versa. Figure 5.1c illustrates a situation in which there is no relationship between the two variables. Moisture content and number of alders vary independently of each other.

In Figure 5.1a the relationship is quite strong, in the sense that knowing the moisture content of a particular quadrat would permit the number of alder trees to be predicted quite accurately, and vice versa. The points on the graph are quite close to a straight line. The relationship shown in Figure 5.1b is rather weaker. The points on the graph are more scattered about a straight line, and predictions of the value of one variable from a knowledge of the other would be less reliable. The third situation (Figure 5.1c) is one of a very weak, or nonexistent, relationship between the two variables. For any particular value of moisture content there is a wide spread of numbers of alders. Predictions based on this graph would be very unreliable.

The two techniques discussed in this chapter provide statistical measures of the strength and direction of a relationship between two variables. The statistical term for such a measure is a *correlation coefficient*. Both the correlation coefficients discussed here can have a value between -1.0 and 1.0. A value of -1.0 indicates a perfect inverse relationship, or perfect *negative correlation*, between the two variables. A value of 1.0 indicates a perfect direct relationship, or perfect *positive correlation*. A complete absence of relationship, or no correlation, is indicated by a coefficient of 0.0.

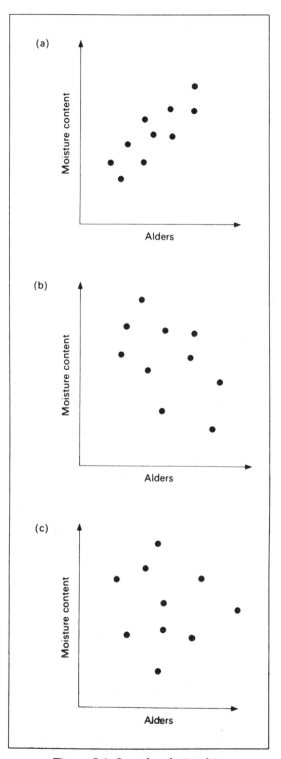

(a)

Moisture content

Alders

(b)

Moisture content

Alders

(c)

Moisture content

Alders

Figure 5.1 Sample relationships

In addition to their purely descriptive use in measuring correlation between two variables, the two coefficients can be used in tests of the significance of a correlation. If the data concern random samples, it is possible to estimate how probable it is that an observed correlation between two variables represents a 'real' correlation in the population from which the samples were taken.

5.1 Product-Moment Correlation

The *product-moment correlation coefficient*, or Pearson's correlation coefficient, is a **parametric** measure of the relationship between two variables. The variables must be measured on an **interval** scale, and the technique assumes that both variables have come from normally distributed populations. This means that if all the individuals in the population were examined the frequency distribution of the values of both variables would be found to be normal. This is a very restricting assumption, which is seldom met by geographical data. Nevertheless product-moment correlation is widely used in geography, to the extent that in geographical literature the term correlation can usually be taken to mean product-moment correlation. It can be argued that the technique is widely misused, in situations where a nonparametric alternative, such as Spearman's rank correlation (Section 5.2), is more appropriate.

Calculation

Table 5.1 contains some data from the heathland study mentioned earlier. Moisture content and number of alder trees have been recorded for each of 10 randomly selected quadrats. The product-moment correlation between the two variables can be calculated using the following equation:

$$r = \frac{\sum xy/n - \bar{x}\bar{y}}{s_x s_y} \qquad (5.1)$$

where r is the product-moment correlation coefficient, x and y refer to the values of the two variables, $\bar{x}$ and $\bar{y}$ are the means of the two variables, and s_x and s_y are the sample standard

Table 5.1 Data for Product-moment Correlation

a	b	c	d	e
x Number of alders	y Moisture content	x^2	y^2	xy
47	21	2209	441	987
35	18	1225	324	630
28	9	784	81	252
27	25	729	625	675
44	27	1936	729	1188
66	38	4356	1444	2508
60	33	3600	1089	1980
75	48	5625	2304	3600
73	49	5329	2401	3577
45	32	2025	1024	1440
$\sum x = 500$ $n = 10$ $\bar{x} = 50$	$\sum y = 300$ $n = 10$ $\bar{y} = 30$	27818 $\bar{x}^2 = 2500$	10462 $\bar{y}^2 = 900$	16837 $\bar{x}\bar{y} = 1500$

deviations of the two variables. As a reminder, the equation for the sample standard deviation is:

$$s_x = \sqrt{\frac{\sum x^2}{n} - \bar{x}^2}$$

Columns c, d and e of table 5.1 show the calculation of the various elements of equation 5.1. For the example data:

$$s_x = \sqrt{\frac{27818}{10} - 2500} = \sqrt{281.8} = 16.79$$

$$s_y = \sqrt{\frac{10462}{10} - 900} = \sqrt{146.2} = 12.09$$

$$r = \frac{\dfrac{16837}{10} - 1500}{16.79 \times 12.09} = \frac{183.7}{203.0} = 0.905$$

Remember that the product-moment correlation coefficient can vary between −1.0 and 1.0, with 0.0 indicating no correlation. A value of 0.905 therefore indicates a very strong positive correlation between soil moisture content and number of alders, on the basis of the sample

data. However, a sample of 10 quadrats is a very small one; it remains to be seen how significant this correlation is.

The reader may come across many alternative equations for the product-moment correlation coefficient. Equation 5.1 is recommended as being the simplest to use for manual calculation. It is **not** the best equation to use as the basis for a computer program (see Mather, 1976, example program 11). It is difficult to give a simple explanation of Equation 5.1. However, it may be helpful to think of the numerator $(\sum xy/n - \bar{x}\bar{y})$ as a measure of the extent to which the two variables vary together, in the sense that an increase in one is mirrored by a proportionate increase or decrease in the other. It is in fact known as the *covariance* of the two variables. A covariance of zero indicates that the two variables do not vary together. A large covariance, either positive or negative, indicates that the two variables do vary together.

The denominator in Equation 5.1 ($s_x s_y$) is a measure of the total variation in the data. In terms of a scatter graph, it is a measure of the spread of the points. The product-moment correlation coefficient can be thought of as the ratio of the joint variation of two variables to the total variation. If the covariance is the same as the

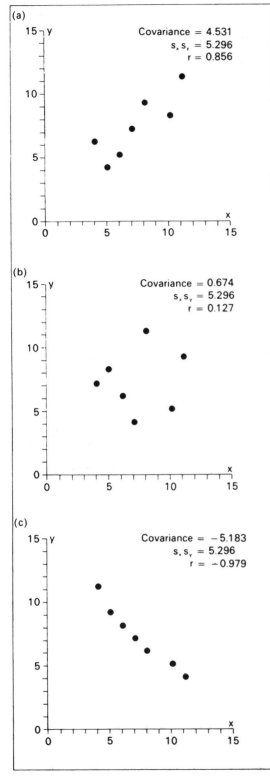

(a)

Covariance = 4.531
$s_x s_y = 5.296$
r = 0.856

(b)

Covariance = 0.674
$s_x s_y = 5.296$
r = 0.127

(c)

Covariance = −5.183
$s_x s_y = 5.296$
r = −0.979

Figure 5.2 Product-moment correlation examples

total variation, the ratio will be 1.0 or −1.0; the total variation will be entirely accounted for by the joint variation of the two variables. If the covariance is small relative to the total variation, the ratio will be close to zero.

This is illustrated in Figure 5.2. For each of the situations shown the total variation ($s_x s_y$) is the same. In each case the same individual values of x and y have been used; only the pairings of x and y values have been changed. The reader can check for himself that the calculations have been carried out correctly. Figure 5.2a shows a situation in which the covariance is quite large relative to the total variation, producing a high correlation coefficient of 0.86. The situation in Figure 5.2b is one in which the covariance is small in relation to the total variation, producing a low correlation coefficient of 0.13. When the covariance is negative, but still large in absolute value compared with the total variation, the correlation coefficient will be close to −1.0. This situation is illustrated in Figure 5.2c (correlation coefficient = −0.98).

In the three situations shown in Figure 5.2 the total variation is constant, i.e. the overall spread of the points is the same in each case. However, the covariance, i.e. the scatter of points about a straight line, is different for each of the situations shown. It is the ratio of these two values which gives the product-moment correlation coefficient.

Exercise 5.1

Calculate the product–moment correlation coefficient for the data given in Table 5.2. The figures are measurements of two variables, rainfall and altitude, for a random sample from the 50 towns in the data matrix (Appendix E, variables K and L). For the purpose of the test, assume that both variables have normal distributions in the population.

Significance

The correlation coefficient can be used simply as a descriptive measure of the degree of correlation between two variables in a sample. This should strictly be termed the *sample correlation coefficient*. If the researcher wishes to know the

Table 5.2 Data for Correlation and Regression Exercises

x (altitude)	y (rainfall)
6.1	9.50
15.5	7.71
23.5	12.29
21.0	9.21
3.7	6.30
0.9	7.20
0.2	7.86
16.7	13.07
0.2	7.71
1.0	8.09

significance of a particular correlation he needs to know how probable it is that the sample correlation coefficient is an accurate estimate of the *population correlation coefficient*. To make the distinction between sample and population coefficients, the sample correlation coefficient is denoted by r, and the population coefficient by ρ (the Greek letter rho).

The null hypothesis of product-moment correlation is that the two sets of measurements (remember that there is one pair of measurements for each individual) are random samples from two independent, normally distributed populations of measurements. This means that if every possible pair of measurements could be made, both variables would be found to be normally distributed. A scatter plot of the pairs of measurements would, however, reveal no correlation (as in Figure 5.1c). Any apparent correlation in the sample data is due to sampling fluctuations. This is equivalent to saying that under the null hypothesis the population correlation coefficient ρ is zero. The fact that the sample coefficient r is not zero is due to chance in the sampling process.

The alternative hypothesis states that there is a correlation between the two variables in the population, i.e. that the population coefficient is **not** zero. As with some of the comparative tests discussed in Chapter 4, the alternative hypothesis can take one of two forms. If it merely specifies that there is a correlation, without specifying a direction of correlation, the test is a two-tailed test. A one-tailed test is used if

the alternative hypothesis specifies a direction of correlation (either positive or negative).

The hypotheses can be summarised in symbolic form:

$$H_0: \rho = 0$$
$$H_1: \rho \neq 0 \quad \text{(two-tailed test)}$$
or $\quad H_1: \rho > 0 \quad \text{(one-tailed test)}$
or $\quad H_1: \rho < 0 \quad \text{(one-tailed test)}$

The null hypothesis situation is illustrated in Figure 5.3. The idea of a probability distribution was introduced in Section 1.3. Figure 5.3 shows a *joint probability distribution* of two variables, x and y, represented as a *probability surface*. The **volume** beneath any part of the surface is proportional to the probability of an individual occurring with those particular values of x and y. Note that the x and y scales are marked off in units of standard deviation. The conversion of interval values into units of standard deviation is explained in Section 1.3. Standardization enables the situation to be generalized to apply to any variables, without needing to worry about the range of values in particular sets of data.

The shaded area in Figure 5.3 lies between 0 and −1 standard deviations on the y scale, and −1 and −2 standard deviations on the x scale. Since both variables have normal distributions, the probability of an individual having a value of

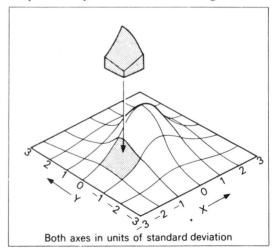

Both axes in units of standard deviation

Figure 5.3 The null hypothesis situation for product-moment correlation: probability surface

x between −1 and −2 units of standard deviations can be calculated as 0.136, using the method explained in section 1.3. Similarly the probability of an individual having a y value between 0 and −1 units of standard deviation can be found to be 0.341. The probability of an individual having an x value between the specified limits **and** a y value between the specified limits is therefore the product of these two probabilities, i.e. $0.136 \times 0.341 = 0.046$. This follows the multiplication rule for probabilities, discussed in Section 1.3.

In this way it is possible to calculate the probability of occurrence of an individual with any particular values of x and y. Figure 5.4 shows, on the basis of these probabilities, the sort of scatter graph predicted by the null hypothesis for a population of 100 points. Again the x and y scales are expressed in units of standard deviation. The null hypothesis, then, is that a survey of the population would reveal a situation like the one shown in Figure 5.4, in which there is no correlation between the two variables. The value of ρ is therefore zero.

Suppose a sample set of data produces a correlation coefficient of 0.8, indicating a strong positive correlation. What is the probability that this sample could have come from a population

in which ρ is zero? Taking into account the size of the sample, and based on the assumption that both variables are normally distributed, it is possible to calculate the sampling distribution of r under the null hypothesis (that $\rho = 0$).

The shape of this sampling distribution is shown in Figure 5.5. It is in fact related mathematically to Student's t distribution, discussed in Section 4.4. Figure 5.5a shows the way in which the sampling distribution of r is used in a test against an alternative hypothesis that there is a correlation between the variables in the population (i.e. that $\rho \neq 0$). This is a *two-tailed test*, since the researcher is concerned with the probability of getting a large value of r, without specifying whether it is positive or negative.

If the alternative hypothesis is that the population coefficient is less than zero $(\rho < 0)$, a *one-tailed test* is appropriate. The researcher is

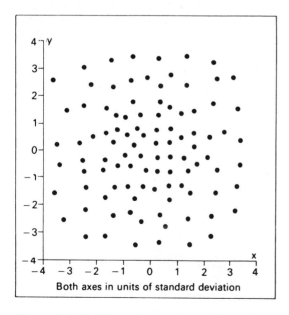

Figure 5.4 *Null hypothesis situation for product-moment correlation: scattergraph*

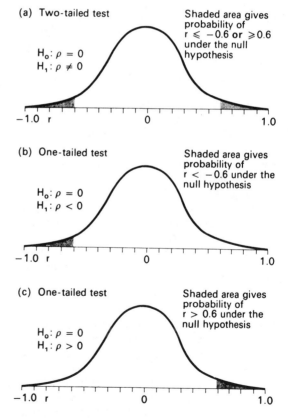

Figure 5.5 *Sampling distribution of the product-moment correlation coefficient*

concerned with the probability of getting a large negative value of r. This is given by the area in the left hand tail of the sampling distribution, as shown in Figure 5.5b. Figure 5.5c illustrates a third possibility: a one-tailed test against an alternative hypothesis that the population coefficient is greater than zero ($\rho > 0$). Again, interest is only in one tail of the sampling distribution, this time the right-hand tail.

The degrees of freedom for a significance tests of the product-moment correlation coefficient are the number of individuals minus two $(n - 2)$. Remember that each individual represents a pair of variable measurements, such as soil moisture content and number of alders. The degree of freedom can therefore also be thought of as the number of pairs of values minus two.

So, for example, for a two-tailed test $(H_1 : \rho \neq 0)$ the critical value of r at the 0.05 significance level with 10 degrees of freedom is 0.576 (from Appendix C8). This means that when the population coefficient is zero $(\rho = 0)$ the probability of a random sample of 12 individuals producing a coefficient as extreme as 0.576 or more $(r \geqslant 0.576$ or $r \leqslant -0.576)$ is 0.05. A value of r greater than or equal to 0.576 (or less than or equal to -0.576) is sufficient to enable the null hypothesis to be rejected in favour of a nondirectional alternative hypothesis at the 0.05 significance level.

For a one-tailed test $(H_1 : \rho > 0)$ the critical value of r at the 0.05 significance level with 10 degrees of freedom is 0.497. When the population coefficient is zero $(\rho = 0)$ the probability of a random sample of 12 individuals producing a coefficient as extreme as 0.497 or more $(r \geqslant 0.497)$ is 0.05. A value of r greater than or equal to 0.497 is sufficient to enable the null hypothesis to be rejected in favour of a directional alternative hypothesis at the 0.05 significance level.

If the alternative hypothesis is that the population coefficient is less than zero $(H_1 : \rho < 0)$ the critical value of r at the 0.05 significance level with 10 degrees of freedom is -0.497. When the population coefficient is zero $(\rho = 0)$ the probability of a random sample of 12 individuals producing a coefficient as extreme as -0.497 or more $(r \leqslant -0.497)$ is 0.05. A value of r less than or equal to -0.497 is sufficient for the null hypothesis to be rejected at the 0.05 significance level. Again, this is a one-tailed test.

For both forms of one-tailed test $(H_1 : \rho > 0,$ and $H_1 : \rho < 0)$ the initial value of r is the same except for a change of sign. It should be noted, however, that the tables of critical values (Appendix C8) give only positive values of r. The reader must be aware that in some circumstances the equivalent negative value of r is the required critical value. It is also important to note that each column of the table of critical values refers to two different significance levels. One is relevant to a two-tailed test, the other to a one-tailed test.

In applying a one-tailed test of correlation it is necessary first of all to establish whether the correlation is in the direction specified in the alternative hypothesis. If the alternative hypothesis is that the population correlation coefficient is negative $(H_1 : \rho < 0)$ the null hypothesis can only be rejected if the sample correlation coefficient r is negative **and** larger in absolute size than the critical value at the chosen significance level.

It is more likely than not that the researcher will have an alternative hypothesis which specifies a direction of correlation. It is difficult to think of many situations, in geography at least, in which he will hypothesize the existence of a correlation without expecting that correlation to be in a particular direction. The one-tailed test is therefore likely to be of more relevance to a geographer. However, it is important for the reader to be clear on the distinction between the two forms of test.

Exercise 5.2

(a) What is the critical value of r for a one-tailed test $(H_1 : \rho < 0)$ with 20 degrees of freedom at the 0.01 significance level?

(b) What is the critical value of r for a two-tailed test $(H : \rho \neq 0)$ with 10 degrees of freedom at the 0.01 significance level?

(c) A random sample survey involving 15 individuals produces a correlation coefficient of -0.48. Is this significant at the 0.05 level (i) using a one-tailed test? (ii) using a two-tailed test?

(d) Is the correlation coefficient calculated for exercise 5.1 significant at the 0.05 level? You will first need to decide whether a one-tailed or two-tailed test is appropriate.

Linearity

Figure 5.6 shows as scatter graphs three different sorts of relationship between variables. The

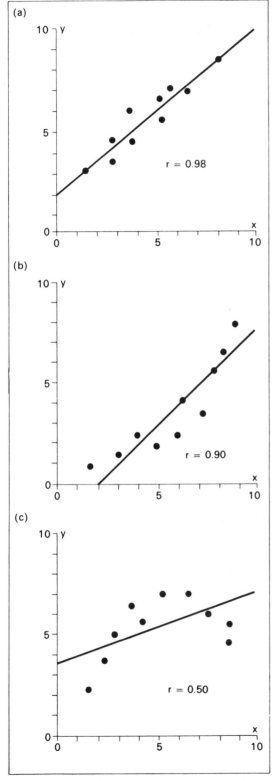

Figure 5.6 Product-moment correlation: linear and nonlinear relationships

value of the product-moment correlation coefficient for each one is also given. The first situation (Figure 5.6a) is one in which there is a clear positive correlation between x and y, and the correlation coefficient of 0.98 supports this visual impression. The second situation produces a correlation coefficient almost as high (0.90), but in the third case the coefficient is only 0.50. However, in both the second and third case, there appears to be a strong relationship between the variables arguably just as strong as in the first case. It would be possible, by eye, to draw a smooth line through the clusters of points in both Figure 5.6b and c. These lines would not, however, be **straight** lines.

The product-moment correlation coefficient measures the extent to which the points on a scatter plot are clustered about a straight line. In Figure 5.6 this straight line has been drawn through each of the scatters of points. The correlation coefficient in each case is a measure of the dispersion of the points about the straight line. In other words, the product moment correlation coefficient is a measure of the **linear** relationship between two variables. Product-moment correlation measures the strength and direction of a relationship between two variables, but not the form of that relationship. The form is assumed to be linear.

More will be said about this topic in the section on regression (Section 6.1). The important point to remember is that the product-moment correlation coefficient is a measure only of the strength and direction of a linear relationship between two variables. It is for this reason that it is essential to draw a scatter graph of the situation before calculating a product moment correlation coefficient. If the pattern of points on the scatter graph suggests that the relationship between the variables is not linear, then the product-moment correlation coefficient should not be applied. There are, however, ways in which non-linear relationships between variables can be transformed into linear relationships. This topic will be discussed in Section 6.2.

5.2 Spearman Rank Correlation

The Spearman rank correlation coefficient is a **nonparametric** measure of the relationship between two sets of **ordinal** (ranked) values. It is a technique which deserves to be more widely used by geographers as an alternative to product moment correlation with its restricting assumptions of two normally distributed variables.

The Spearman test can be applied to data which are inherently ordinal, such as preference data, or to interval data converted to ranked form. For the sake of simplicity the calculation of the Spearman rank correlation coefficient will initially be demonstrated in connection with the first type of data.

Calculation

Four people are asked to rank ten towns in order of preference as retirement homes. The town they prefer most is ranked 1, the least popular town is ranked 10. Table 5.3 shows these four sets of rankings. The towns have been listed in alphabetical order, but this is of no relevance to the calculation of the correlation coefficient.

Rank correlation provides a means of calculating the degree of correspondence between any two sets of rankings. The equation for the Spearman rank correlation coefficient is:

$$r_s = 1 - \frac{6 \sum d^2}{n^3 - n} \qquad (5.2)$$

Where r_s is the Spearman rank correlation coefficient, d is the difference in ranking for each item, $\sum d^2$ means that the differences must

be squared and then summed, and n is the number of items ranked, i.e. the number of pairs of rankings.

The last two columns of table 5.3 show the calculation of d, d^2 and $\sum d^2$ for the first two sets of rankings (Peter and Jane). The value of the rank correlation coefficient can now be calculated:

$$r_s = 1 - \frac{6 \times 34}{10^3 - 10} = 1 - \frac{204}{990} = 1 - 0.206$$

$$= 0.794$$

Like the product-moment correlation coefficient, the Spearman coefficient can have a value between -1.0 indicating perfect negative correlation between the two sets of rankings, and 1.0, indicating perfect positive correlation. A value of 0.0 indicates an absence of correlation.

In terms of preferences for retirement towns a coefficient of -1.0 would mean that two people ranked the towns in exactly opposite orders. Two identical sets of rankings would produce a coefficient of 1.0. A coefficient of 0.0 indicates that two people's rankings are neither the same nor opposite. Knowing one person's ranking for a particular town would give no indication of the other person's ranking in this case.

Exercise 5.3

Calculate the degree of correlation between the rankings of Michael and John (in Table 5.3).

Table 5.3 Preferences for Retirement Homes

Towns	Rankings				Differences between ranks (Peter and Jane)	
	Peter	Jane	Michael	John	d	d^2
Aberdeen	10	10	1	6	0	0
Birmingham	7	8	4	10	−1	1
Bournemouth	1	2	8	2.5	−1	1
Bristol	5	5	7	4	0	0
Hastings	2	1	9	1	1	1
Llandudno	9	4	3	7	5	25
London	4	6	5	8	−2	4
Manchester	8	9	2	5	−1	1
Norwich	6	7	6	9	−1	1
Torquay	3	3	10	2.5	0	0
					$\sum d^2 = 34$	

Note that the usual convention for tied ranks has been adopted in the case of John's rankings of Bournemouth and Torquay.

So far rank correlation has only been applied to inherently ranked data. To demonstrate how the technique can be applied to interval measurements, the data concerning moisture content and alder trees given in Section 5.1 will be considered. These data are repeated in the first two columns of Table 5.4. The third column contains the rankings of the number of alders. They have been ranked from largest to smallest, i.e. 75 becomes 1 and 27 becomes 10. The rankings of the second variable, moisture content, are given in the fourth column. The differences between the ranked values for each sample point and the squares of these differences are also given in table 5.4.

The rank correlation coefficient can now be calculated using equation 5.2:

$$r_s = 1 - \frac{6 \sum d^2}{n^3 - n} = 1 - \frac{6 \times 24}{10^3 - 10}$$

$$= 1 - \frac{144}{990} = 0.855.$$

For comparison, the value of the product-moment correlation coefficient for the same data in interval form is 0.905 (Section 5.1). It must be remembered that the process of converting from interval to ordinal measurements is bound to result in some loss of information.

There is no reason to expect that rank correlation should produce the same result as product-moment correlation when the two techniques are applied to a common data set. Rank correlation may well be a more reliable measure in many instances, since it does not depend on any, possibly unwarranted, assumptions about the frequency distributions of the variables.

Significance

In the first example in this section the data do not come from a sample study. The preferences of Peter and Jane are not sample measurements, they simply **are** the preferences of these two people. A significance test in this situation would therefore have no meaning. In the second example, however, a significance test is possible, since the data concern measurements made at a random sample of quadrats. The null hypothesis is that there is no rank correlation between moisture content and number of alders in the population. In other words the population rank correlation coefficient (denoted by ρ_s) is zero. The fact that the sample rank correlation coefficient (r_s) is not zero is due to chance in the sampling process.

The alternative hypothesis may be either directional or non-directional. A non-directional test ($H_1: \rho_s \neq 0$) involves two tails of the sampling distribution of r_s. A directional test

Table 5.4 Original and Ranked Data for Spearman Rank Correlation

x Number of alders	y Moisture content	Ranked values		Rank differences	
		r_x	r_y	d	d^2
47	21	5	8	−3	9
35	18	8	9	−1	1
28	9	9	10	−1	1
27	25	10	7	3	9
44	27	7	6	1	1
66	38	3	3	0	0
60	33	4	4	0	0
75	48	1	2	−1	1
73	49	2	1	1	1
45	32	6	5	1	1
n = 10				$\sum d^2 = 24$	

($H_1: \rho_s < 0$ or $H: \rho_s > 0$) is one-tailed. The degrees of freedom for a significance test of r_s are the number of pairs of ranked values, n. The shape of the sampling distribution of r_s is very similar to that of the product-moment correlation distribution shown in Figure 5.5. The explanation of one-tailed and two-tailed tests applied to product-moment correlation (Section 5.1) applies equally to rank correlation. That explanation will not be repeated here. The important point to remember is that critical values of r_s for a one-tailed test are different from those for an equivalent two-tailed test.

The rank correlation between moisture content and number of alders is 0.855, and the degrees of freedom are 10 (the number of pairs of rankings). Assuming a significance level of 0.05, the critical value of r_s for a one-tailed test is found from tables Appendix C9) to be 0.564. For a two-tailed test at the same significance level the critical value is 0.648. This means that there is a probability of 0.05 of getting a sample rank correlation of 0.564 or more when the population correlation is zero. This is a one-tailed test ($H_1: \rho_s > 0$). There is a probability of 0.05 of getting a sample rank correlation of

0.648 or more, or −0.648 or less, when the population correlation is zero. This is a two-tailed test ($H_1: \rho_s \neq 0$).

For the correlation between moisture content and alders the alternative hypothesis is likely to be directional: that there is a positive correlation between the two variables in the population. A one-tailed test is therefore appropriate. The calculated vale of r_s is positive **and** greater than the critical value for a one-tailed test at the chosen significance level of 0.05. The null hypothesis can therefore be rejected, and the conclusion can be drawn that there is a significant rank correlation between moisture content and numbers of alders at the 0.05 level.

As with some other non-parametric tests, it is possible to construct the sampling distribution of r_s without recourse to statistical theory. The sampling distribution of r_s with three degrees of freedom can be constructed by considering all the different pairings of three ranks. These pairings are listed in Table 5.5a. The value of r_s produced by each combination is also given. It can be seen that out of the 6 possible pairings, 1 produces an r_s of −1.0, 2 produce an r_s of −0.5, 2 produce an r_s of 0.5 and 1 produces an r_s of

Table 5.5

(a) All Possible Pairings of Ranks of Three Items in the Spearman Rank Correlation Test

Pairing (1/2 means '1 with 2')			Value of r_s from this pairing
1/1	2/2	3/3	1.0
1/1	2/3	3/2	0.5
1/2	2/1	3/3	0.5
1/2	2/3	3/1	−0.5
1/3	2/1	3/2	−0.5
1/3	2/2	3/1	−1.0

(b) Sampling Distribution of r_s with 3 Degrees of Freedom

r_s	probability
−1.0	0.1667 (1/6)
−0.5	0.3333 (2/6)
0.5	0.3333 (2/6)
1.0	0.1667 (1/6)

1.0. Under the null hypothesis, that the rankings are uncorrelated in the population from which the sample is taken, all these pairings are equally probable. It is therefore possible to construct a probability distribution from the data given in Table 5.5a. This probability distribution is shown in Table 5.5b. If a sample contains only three pairs of ranked values it is quite easy to obtain a value of r_s as large as 1.0 when the correlation between the variables in the population is zero. As sample size increases, it becomes less and less easy to obtain a large value of r_s, whether negative or positive, when ρ_s is zero.

Exercise 5.4

Calculate the Spearman rank correlation coefficient for the sample data given in Table 5.2. Is this correlation significant at the 0.01 level? You will first need to decide whether a one-tailed or a two-tailed test is appropriate.

Correction for Tied Ranks

As with other statistical tests for ordinal data, the presence of tied ranks slightly alters the sampling distribution. If there is only a small proportion of tied ranks the effect is negligible. However, if the proportion of ties is large, the following equation should be used instead of Equation 5.2:

$$r_s = \frac{(A-B)+(A-C)-\sum d^2}{2\sqrt{(A-B)(A-C)}} \quad (5.3)$$

where $\sum d^2$ is the sum of the squares of the differences between each x and y ranking (as in Equation 5.2). Also:

$$A = \frac{n^3 - n}{12}$$

where n is the number of pairs of rankings.

$$B = \sum\left(\frac{t_x^3 - t_x}{12}\right)$$

$$C = \sum\left(\frac{t_y^3 - t_y}{12}\right)$$

where t_x is the number of values of variable x tying at a given rank, and t_y is the same for variable y. The summation sign $\sum$ means that the value of $((t_x^3 - t_x)/12)$ is calculated for each group of ties at a particular rank, and these values are then summed.

The calculation of this correction factor can be demonstrated by applying it to Exercise 5.3 (Michael and John's rankings of towns). There are no ties in variable x (Michael's rankings) so the value of B for Equation 5.3 is 0.0. In John's rankings (variable y) there is just one group of ties: Bournemouth and Torquay have both been given a ranking of 2.5. The value of C for Equation 5.3 is therefore:

$$\frac{2^3 - 2}{12} = \frac{8-2}{12} = \frac{6}{12} = 0.5$$

The value of $\sum d^2$ should have already been calculated for Exercise 5.3 as 263.5. The number of pairs of rankings n is 10, so A can be calculated as:

$$\frac{10^3 - 10}{12} = \frac{990}{12} = 82.5$$

All these values can be substituted into Equation 5.3 in order to calculate r_s corrected for ties:

$$r_s = \frac{(82.5-0.0)+(82.5-0.5)-263.5}{2\sqrt{(82.5-0.0)(82.5-0.5)}}$$

$$= \frac{82.5+82.0-263.5}{2\sqrt{82.5\times82.0}}$$

$$= \frac{164.5-263.5}{2\sqrt{6765}} = \frac{-99.0}{2\times82.25} = -0.602$$

The reader can see that this value of r_s is very little different from the value calculated without the correction for ties (−0.597). In fact, when rounded to two places of decimals, the two values are identical. It should also be noted that the effect of the correction for ties is to increase the value of r_s, and therefore make it easier for the null hypothesis to be rejected. In other words, if the correction is not applied, the significance test errs on the side of caution. For these reasons it is probably unnecessary to use the correction unless there is a very large proportion of tied ranks in the data.

5.3 Computer Programs

Program 9 — Product–Moment Correlation

This program calculates Pearson's product-moment correlation coefficient for the relationship between two variables.

Program Listing

```
10 REM - PRODUCT-MOMENT CORRELATION
20 PRINT"-------------------------"
30 PRINT"PRODUCT-MOMENT CORRELATION"
40 PRINT"-------------------------"
50 PRINT
60 INPUT"Number of values of each variable",NX
70 N=NX+NX
80 DIM X(N)
90 PRINT
100 PRINT"Now type in values of X:"
110 PRINT"(Mistakes can be corrected later)"
120 FOR I=1 TO NX
130 PRINT I;" ";
140 INPUT X(I)
150 NEXT I
160 PRINT
170 PRINT"Now type in values of Y:"
180 PRINT"(Mistakes can be corrected later)"
190 FOR I=NX+1 TO N
200 PRINT I-NX;" ";
210 INPUT X(I)
220 NEXT I
230 N1=1
240 N2=NX
250 V$="X"
260 GOSUB 560
270 N1=NX+1
280 N2=N
290 V$="Y"
300 GOSUB 560
310 FOR I=1 TO NX
320 XM=XM+X(I)
330 YM=YM+X(I+NX)
340 SX=SX+X(I)^2
350 SY=SY+X(I+NX)^2
360 XY=XY+X(I)*X(I+NX)
370 NEXT I
380 XM=XM/NX
390 YM=YM/NX
400 SX=SQR(SX/NX-XM^2)
410 SY=SQR(SY/NX-YM^2)
420 R=(XY/NX-XM*YM)/(SX*SY)
430 PRINT
440 PRINT"-------------------------"
450 PRINT"PRODUCT-MOMENT CORRELATION"
460 PRINT"-------------------------"
470 PRINT
480 PRINT"Mean of X = ";XM
490 PRINT"Standard deviation of X = ";SX
500 PRINT"Mean of Y = ";YM
510 PRINT"Standard deviation of Y = ";SY
520 PRINT"r = ";R
530 DF=NX-2
540 PRINT"with ";DF;" degrees of freedom"
550 STOP
560 REM - DATA CHECKING ROUTINE
570 PRINT
580 PRINT"Do you want to check/correct ";V$;
590 INPUT" values (Y/N)";Q$
600 IF Q$="N" GOTO 710
610 IF Q$<>"Y" GOTO 580
620 K=0
630 PRINT"Press RETURN if value is correct";
640 PRINT" - otherwise type correct value"
650 FOR I=N1 TO N2
660 K=K+1
670 PRINT K;" ";X(I);
680 INPUT A$
690 IF A$<>"" X(I)=VAL(A$)
700 NEXT I
710 RETURN
```

Example Output

```
--------------------------
PRODUCT-MOMENT CORRELATION
--------------------------

Number of values of each variable?10

Now type in values of X:
(Mistakes can be corrected later)
        1 ?47
        2 ?35
        3 ?28
        4 ?27
        5 ?44
        6 ?66
        7 ?60
        8 ?75
        9 ?73
       10 ?45

Now type in values of Y:
(Mistakes can be corrected later)
        1 ?21
        2 ?18
        3 ?9
        4 ?25
        5 ?27
        6 ?38
        7 ?33
        8 ?48
        9 ?49
       10 ?32

Do you want to check/correct X values (Y/N)?N

Do you want to check/correct Y values (Y/N)?N

--------------------------
PRODUCT-MOMENT CORRELATION
--------------------------

Mean of X = 50
Standard deviation of X = 16.7868997
Mean of Y = 30
Standard deviation of Y = 12.0913192
r = 0.905034167
with 8 degrees of freedom

STOP at line 550
```

The data used are from the worked example in
the text (Table 5.1).

Program 10 — Spearman Rank Correlation

This program calculates Spearman's rank correlation coefficient for the relationship between
two variables.

Program Listing

```
10 REM - SPEARMAN RANK CORRELATION
20 PRINT"-----------------------"
30 PRINT"SPEARMAN RANK CORRELATION"
40 PRINT"-----------------------"
50 PRINT
60 INPUT"Number of values of each variable",NX
70 N=NX+NX
80 DIM X(N),N(N),R(N)
90 PRINT
100 PRINT"Now type in values of X:"
110 PRINT"(mistakes can be corrected later)"
120 FOR I=1 TO NX
130 PRINTI;" ";
140 INPUT X(I)
150 N(I)=I
160 NEXT I
170 PRINT
180 PRINT"Now type in values of Y:"
190 PRINT"(mistakes can be corrected later)"
200 FOR I=NX+1 TO N
210 PRINT I-NX;" ";
220 INPUT X(I)
230 N(I)=I
240 NEXT I
250 N1=1
260 N2=NX
270 V$="X"
280 GOSUB 940
290 N1=NX+1
300 N2=N
310 V$="Y"
320 GOSUB 940
330 A=(NX^3-NX)/12
340 N1=1
350 N2=NX
360 GOSUB 570
370 B=A-T
380 N1=NX+1
390 N2=N
400 GOSUB 570
410 C=A-T
420 A=(NX^3-NX)/12
430 D=0
440 FOR I=1 TO NX
450 D=D+(R(I)-(R(NX+I)-NX))^2
460 NEXT I
470 RS=(B+C-D)/(2*SQR(B*C))
480 PRINT
490 PRINT"-----------------------"
500 PRINT"SPEARMAN RANK CORRELATION"
510 PRINT"-----------------------"
520 PRINT
530 PRINT"Rs = ";RS
540 PRINT"(corrected for ties)"
550 PRINT"with ";NX;" degrees of freedom"
```

```
560 STOP
570 REM - BUBBLE SORT RANKING ROUTINE
580 Z=0
590 FOR I=N1 TO N2-1
600 J=I+1
610 IF X(I)<X(J) GOTO 710
620 IF X(I)>X(J) GOTO 640
630 GOTO 710
640 D=X(J)
650 DN=N(J)
660 N(J)=N(I)
670 X(J)=X(I)
680 N(I)=DN
690 X(I)=D
700 Z=1
710 NEXT I
720 IF Z=1 GOTO 580
730 REM - CHECK FOR TIED RANKS
740 R(N(N2))=N2
750 NT=0
760 T=0
770 FOR I=N1 TO N2-1
780 FOR J=I+1 TO N2
790 IF X(I)<>X(J) GOTO 820
800 NT=NT+1
810 NEXT J
820 IF NT<>0 GOTO 860
830 R(N(I))=I
840 R(N(J))=J
850 GOTO 920
860 FOR K=I TO J-1
870 R(N(K))=(I+J-1)/2
880 NEXT K
890 T=T+((NT+1)^3-(NT+1))/12
900 I=I+NT
910 NT=0
920 NEXT I
930 RETURN
940 REM - DATA CHECKING ROUTINE
950 PRINT
960 PRINT"Do you want to check/correct ";V$;
970 INPUT" values (Y/N)",Q$
980 IF Q$="N" GOTO 1090
990 IF Q$<>"Y" GOTO 960
1000 K=0
1010 PRINT"Press RETURN if value is correct";
1020 PRINT" - otherwise type correct value"
1030 FOR I=N1 TO N2
1040 K=K+1
1050 PRINT K;" ";X(I);
1060 INPUT A$
1070 IF A$<>"" X(I)=VAL(A$)
1080 NEXT I
1090 RETURN
```

Example Output

```
-------------------------
SPEARMAN RANK CORRELATION
-------------------------

Number of values of each variable?10

Now type in values of X:
(mistakes can be corrected later)
        1  ?1
        2  ?4
        3  ?8
        4  ?7
        5  ?9
        6  ?3
        7  ?5
        8  ?2
        9  ?6
       10  ?10

Now type in values of Y:
(mistakes can be corrected later)
        1  ?6
        2  ?10
        3  ?2.5
        4  ?4
        5  ?1
        6  ?7
        7  ?8
        8  ?5
        9  ?9
       10  ?2.5

Do you want to check/correct X values (Y/N)?N

Do you want to check/correct Y values (Y/N)?N

-------------------------
SPEARMAN RANK CORRELATION
-------------------------

Rs = -0.601826488
(corrected for ties)
with 10 degrees of freedom

STOP at line 560
```

The data used are from Exercise 5.3 (Table 5.3). As it happens, these data are already ranks, but the program will handle interval data equally well. In fact it re-ranks the ranked data in this case! The correction for tied ranks (Equation 5.3) is applied when necessary.

6 Trends

Chapter 5 was concerned with techniques for measuring the strength and direction of a relationship between two variables. This chapter deals with some ways of measuring the form of a relationship between two variables, or, more precisely, of measuring the *dependence* of one variable on another. These sorts of techniques are necessary for prediction.

Figure 6.1 is a graph illustrating the relationship between temperature in degrees Fahrenheit and degrees centigrade. Only two points are marked on the graph, the boiling point of water (212°F, 100°C) and the freezing point of water 32°F, 0°C). Since there is a perfect positive linear relationship (correlation +1.0) between temperatures measured on the two scales, all other temperature points will lie along the straight line illustrated. For example, 30°C can be correctly translated into 86°F, as shown on the graph.

It is also possible to express this relationship in the form of an equation:

$$y = 32 + 1.8x \qquad (6.1)$$

where y is degrees Fahrenheit and x is degrees centigrade. By substituting a value of °C into this equation it is possible to calculate the equivalent Fahrenheit temperature. For example, if $x = 100$, $y = 32 + 1.8 \times 100 = 32 + 180 = 212$. Such an equation is rather easier to use, and more accurate than the graph when converting from centigrade to Fahrenheit.

The purpose of many of the methods discussed in this chapter is to provide a concise description of the form of a relationship between two variables, preferably as an equation. There are, however, two major practical problems to be faced:

(1) The relationship between most geographical variables is not perfect.

(2) The relationship between many geographical variables is not linear.

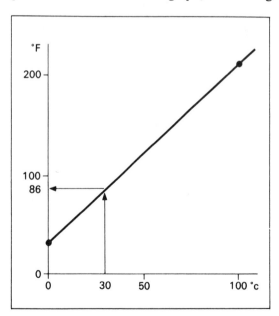

Figure 6.1 Relationship between Fahrenheit and centigrade temperatures

Imperfect Relationships

The graph in Figure 6.2 illustrates the relationship between moisture content and alders

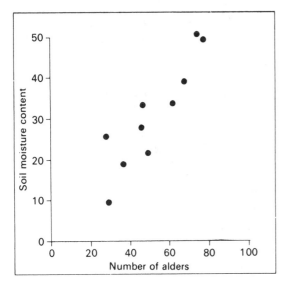

Figure 6.2 Relationship between moisture content and alders

through a set of points on a graph. Each of these will almost certainly produce a line which has a different equation, and consequently predictions made on the basis of each of these lines will also differ. Of equal importance is the fact that these various methods make different assumptions about the nature of the data to which they are applied.

Nonlinear Relationships

It was suggested in the previous chapter that many relationships between geographical variables are not linear. They cannot be adequately summarized by a straight line on a graph. Figure 6.4 illustrates a common form of nonlinear relationship in geography. The data are from a hypothetical sample survey of shopping behaviour patterns in a large city. It can be seen that frequency of shopping in the city centre falls off very sharply over the first one or two kilometres, but then declines much less rapidly with distance. The 'form' of this relationship can best be described by a curve (see Figure 6.4).

Figure 6.5 illustrates another sort of nonlinear relationship. In this case there is a *cyclical* pattern of variation in the data. This sort of pattern can often be found when a geographical variable is measured over time. For example, temperature measured at a particular location can be expected to show this sort of variation, both hour by hour through the day, and through the year with the changing seasons. A set of data measured over time in this way is known as a *time series*, and a specialized branch of statistics has developed, particularly amongst economists, for the analysis of time series.

Finding an equation to describe a nonlinear relationship is a rather more complex problem than dealing with linear relationships. In this chapter only the more elementary methods will be discussed.

discussed in an earlier example (Section 5.1, Table 5.1). Clearly it is not possible to draw a straight line on the graph which will pass through all the points. The problem, then, is to draw a straight line which passes as close to as many of the points as possible. This can be called a *best fit* straight line.

A number of alternative criteria of 'goodness of fit' can be suggested, and some of these are illustrated in Figure 6.3. The line in Figure 6.3a, is drawn in such a way that the sum of the distances from the points to the line, measured at right angles to the line, is as small as possible. The line in Figure 6.3b is known as the *reduced major axis* line. It is drawn in such a way that the total area of all the triangles formed between the points and the line is as small as possible. Figure 6.3c illustrates a line drawn so as to minimize the sum of the distances from the points to the line, measured parallel to the vertical axis. A fourth alternative, illustrated in Figure 6.3d, is a line which minimizes the sum of the squares of the distances from the points to the line, measured parallel to the vertical axis. This line is known as the *least-squares* line, and it is in fact the best fit line most commonly used in geographical studies.

The important point at this stage is that there is more than one way to draw a best fit line

6.1 Simple Linear Regression

It has already been pointed out that there are a number of alternative criteria for fitting a straight line to a set of data. The advantages of the least-squares approach are that it provides a unique solution, it is relatively simple to apply

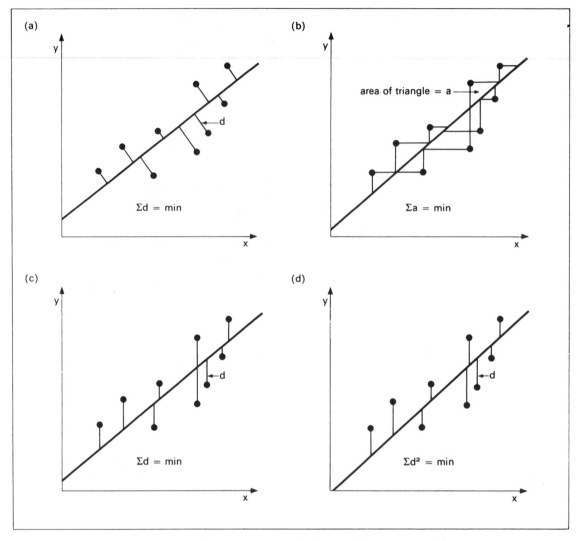

Figure 6.3 Alternative criteria for fitting a line to a set of data

and it possesses certain desirable statistical properties. *Linear regression* is the name conventionally given to a technique for calculating the equation of a least-squares line.

Linear regression is widely used in geographical research. It is however, a technique which it is very easy to misuse. Some of the assumptions involved in the application of linear regression are highly restrictive. These will be discussed later. First it is necessary to make clear the aims of the technique.

The Rationale of Regression

The end product of regression analysis is an equation such as the one mentioned earlier for predicting Fahrenheit temperatures from centigrade temperatures.

This can be written in a general form as:

$$\hat{y} = a + bx \qquad (6.2)$$

where $\hat{y}$ is the predicted value of the *dependent*

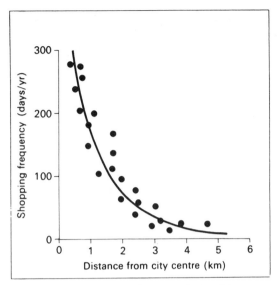

Figure 6.4 A nonlinear relationship

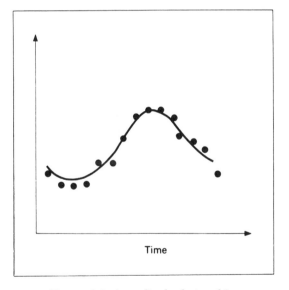

Figure 6.5 A cyclical relationship

variable, x is the value of the *independent vari-able*, and a and b are *regression coefficients*.

The dependent (or regressor) variable is the variable whose values are to be predicted, or explained, given values of the independent (or predictor) variable. It is important to be absolutely clear which is the dependent variable, and which the independent variable in regression.

The assumption embodied in a regression equation is that the values of the dependent variable depend on the values of the independent variable. There is a suggestion of cause and effect. Throughout this chapter the dependent variable will be referred to as y and the independent variable as x. The essential part of regression analysis is the calculation of the regression coefficients a and b, using the following equations:

$$b = \frac{\sum xy - n\bar{x}\bar{y}}{\sum x^2 - n\bar{x}^2} \qquad (6.3)$$

$$a = \bar{y} - b\bar{x} \qquad (6.4)$$

where x refers to the values of the independent variable, and y to the values of the dependent variable, $\bar{x}$ and $\bar{y}$ are the respective means of the two sets of values, and n is the number of pairs of measurements (the number of points on the graph).

From the data on moisture content and number of alders it is possible to calculate the coefficients of the regression equation. In this case the dependent variable is the moisture content and the independent variable is the number of alders. The aim is to produce a regression equation for predicting the moisture content of a quadrat from the number of alders found in it. This is known as the regression of moisture content on number of alders. Such an equation would enable a count of the number of alder trees to act as an indicator of the moisture content of the soil in a particular area.

The basic data are given in Table 6.1, together with the calculation of some of the elements of Equations 6.3 and 6.4. The regression coefficients can now be calculated:

$$b = \frac{\sum xy - n\bar{x}\bar{y}}{\sum x^2 - n\bar{x}^2} = \frac{16837 - (10 \times 50 \times 30)}{2718 - (10 \times 50^2)}$$

$$= \frac{16837 - 15000}{27818 - 25000} = \frac{1837}{2818} = 0.652$$

$$a = \bar{y} - b\bar{x} = 30 - (0.652 \times 50)$$

$$= 30 - 32.6 = -2.6$$

These coefficients can be inserted into a regression equation (Equation 6.2). In this case:

$$\hat{y} = -2.6 + 0.652x \qquad (6.5)$$

Table 6.1 Data for Regression of Moisture Content on Alders

x Number of alders	y Moisture content	x^2	xy
47	21	2209	987
35	18	1225	630
28	9	784	252
27	25	729	675
44	27	1936	1188
66	38	4356	2508
60	33	3600	1980
75	48	5625	3600
73	49	5329	3577
45	32	2025	1440
$\sum x = 500$ n = 10 $\bar{x} = 50$	$\sum y = 300$ n = 10 $\bar{y} = 30$	$\sum x^2 = 27818$	$\sum xy = 16837$

From this equation it is possible to predict the moisture content, ŷ, for any particular number of alders (x). For example, if the number of alders found in a quadrat is 50, the moisture content should be $-2.6 + (0.652 \times 50) = 30.0$. Similarly, a quadrat containing 25 alders should have a moisture content of $-2.6 + (0.652 \times 25) = 13.7$. The least-squares regression line must, therefore, pass through the points on the graph where x = 50 and y = 30.0, and where x = 25 and y = 13.7. Figure 6.6 shows this regression line drawn through the graph of moisture content against alders.

The reader should note at this stage that the least-squares line for the regression of x on y is **not** the same as the least-squares line for the regression of y on x, unless there is a perfect linear relationship between two variables. So, for example, it is **not** possible to re-arrange Equation 6.5 algebraically to produce an equation for predicting x from y. Also, because of the assumptions of the regression technique, discussed later in this section, regression is usually only a one way process. In other words, whilst it may be valid to predict values of y from values of x, it is very unlikely that it will also be valid to attempt to predict values of x from values of y. Dependent and independent variable are not usually interchangeable.

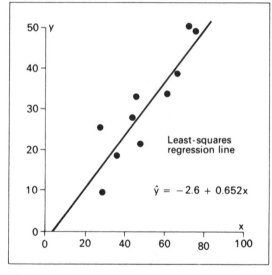

Figure 6.6 Regression of moisture content on number of alders

Exercise 6.1

Calculate the regression equation for the regression of y on x (i.e. an equation for predicting y from x) using the data given in Table 5.2.

Assumptions

Regression relies on more assumptions than any other technique discussed in this book. Yet, frequently regression is applied in geography and elsewhere with an apparent disregard for, or even ignorance of, the assumptions. For this reason they will be discussed in some detail here. If the assumptions are not fairly well satisfied, inferences made from a regression may be invalid, although the regression equation may still be of value in describing the relationship between two variables.

There are five major assumptions:

(1) The independent variable x is not a set of sample values. Strictly speaking the values of x should be fixed values chosen by the researcher, and not sample measurements. If this is not the case, as in the example used here, then the values of the independent variable must at least have been measured with a negligible amount of error.

(2) The values of the dependent variable are normally distributed.

(3) The variance of the dependent variable is constant for all values of the independent variable.

(4) The value of the *residuals* (the differences between the observed and predicted values of the dependent variable) have a normal distribution.

(5) The values of the residuals are independent of each other, i.e. they are randomly arranged along the regression line.

Since these assumptions are often ignored in geographical applications of regression, it is worth emphasizing them by considering a hypothetical situation in which they are all satisfied.

In a particular lake, samples of water are taken at depth intervals of one metre, and the salinity of each sample is measured. Figure 6.7a is a graph of the data obtained, with the least squares regression line drawn through them. The independent variable in this case is depth, which is not a sample variable, but one which has been fixed at certain specified values. It is not subject to sampling error (since a metre is always a metre) and can be measured with a high degree of accuracy. Assumption 1 is therefore satisfied.

The dependent variable, salinity, is a series of sample measurements, which are subject to sample variation. Taking a sample of water at each depth implies selecting a particular 'lump' of water. Since the water is constantly in motion the selection of any particular 'lump' is a chance happening. Assumption 2 means that if the researcher were to take, say, 1000 samples at any one depth, and produce a frequency distribution of the salinity values from these samples, it would be a normal distribution.

Assumption 3 means that if this same process were repeated at each depth, the shapes of all the frequency distributions would be identical. There would however, be a difference between the means of these frequency distributions. In fact the least-squares technique assumes that the means of all the distributions lie exactly along the regression line. Figure 6.7a illustrates assumptions 2 and 3 diagramatically. Figure 6.7b shows the residuals, the deviations of the observed from the expected values of the dependent variable. For assumption 4 to be satisfied, the frequency distribution of the residuals should be normal. According to assumption 5, there should be no pattern in the residuals. They should have all the characteristics of a sequence of random numbers.

So far, regression has been discussed as a method for calculating an equation which can be used as a concise description of form of a relationship between two variables. If the data relate to a sample study, as they frequently do, the researcher needs to know two things:

(1) How well the regression line fits the data.

(2) How accurate any predictions based on the regression, are likely to be.

Goodness of Fit

The equation linking Fahrenheit and centigrade temperatures (Equation 6.1) is clearly a complete explanation of the relationship between the two temperature scales. Equation 6.5, on the other hand, does not provide a complete explanation of the relationship between moisture content and number of alders. The regression line, shown in Figure 6.6, does not pass through any of the observed points. In other words, the predicted value of the dependent variable ($\hat{y}$) is different from the

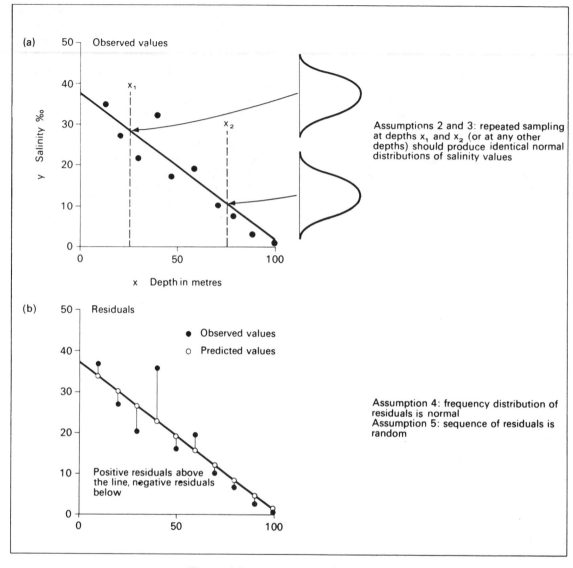

Figure 6.7 Assumptions of regression

observed value (y) for any particular observed value of x. In Figure 6.8 the predicted values of y, obtained from applying Equation 6.5, are marked by open circles, so that their deviations from the observed values, marked by dots, can be seen.

These deviations, given by the observed value of y minus the predicted value $(y - \hat{y})$, are known as the *residuals* from the regression. Clearly

then these residuals are small; the regression line is a good fit. This is the basis of one method for calculating the goodness of fit of a regression line.

The standard method of measuring the goodness of fit of a regression is to calculate the extent to which the regression accounts for the variation in the observed values of the dependent variable. This can be quantified

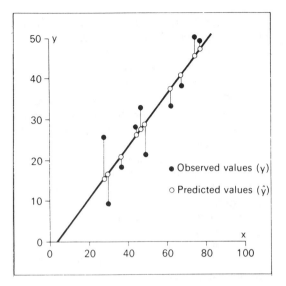

Figure 6.8 Residuals from regression of moisture content on alders

by calculating the variance of the observed values of y using the equation given in Section 2.2:

$$s^2 = \frac{\sum x^2}{n} - \bar{x}^2$$

For the observed values of moisture content:

$$s_y^2 = \frac{\sum y^2}{n} - \bar{y}^2 = \frac{10462}{10} - 30^2$$

$$= 1046.2 - 900 = 146.2$$

The calculation of observed, or total, variance (s_y^2) is given in Table 6.2, and shown diagrammatically in Figure 6.9a. To find out how much of this variation is accounted for by the regression, the variance of the predicted values of the dependent variable can also be calculated. The predicted values ($\hat{y}$) and their squares ($\hat{y}^2$) are

Table 6.2 Calculation of Total Variance and Regression Variance for the Regression of Moisture Content on Alders

| Observed values | | | Predicted values | | Residuals (y − ŷ) | |
x	y	y^2	$\hat{y}$	$\hat{y}^2$	e	e^2
27	25	625	15.00	225.00	10.00	100.00
28	9	81	15.66	245.24	−6.66	44.36
35	18	324	20.22	408.85	−2.22	4.93
44	27	729	26.09	680.69	0.91	0.83
45	32	1024	26.74	715.03	5.26	27.67
47	21	441	28.04	786.24	−7.04	49.56
60	33	1089	36.52	1333.71	−3.52	12.39
66	38	1444	40.43	1634.58	−2.43	5.90
73	49	2401	45.00	2025.00	4.00	16.00
75	48	2304	46.30	2143.69	1.70	2.89
300	10462		300.00	10198.03	0.00	

Total variance: $s_y^2 = \dfrac{\sum y^2}{n} - \bar{y}^2 = \dfrac{10462}{10} - 30^2 = 1046.2 - 900 = 146.2$

Regression variance: $s_{\hat{y}}^2 = \dfrac{\sum \hat{y}^2}{n} - \bar{y}^2 = \dfrac{10198.03}{10} - 30^2 = 1019.803 - 900$

$$= 119.8$$

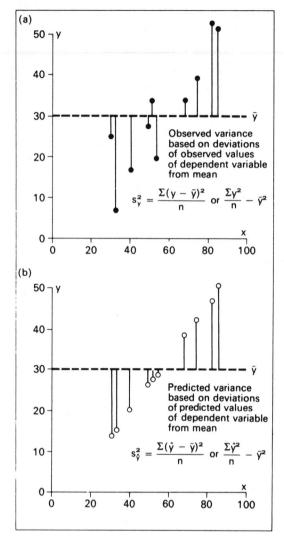

Figure 6.9 Observed and predicted variance in regression analysis

also given in Table 6.2. The predicted, or regression, variance can now be calculated:

$$s_{\hat{y}}^2 = \frac{\sum \hat{y}^2}{n} - \bar{y}^2 = \frac{10198.03}{10} - 30^2$$

$$= 1019.803 - 900 = 119.8$$

This calculation is shown diagrammatically in Figure 6.9b.

The ratio between these two variances provides a measure of the goodness of fit of a regression. This ratio is known as the *coefficient of determination*, which has the symbol r^2:

$$r^2 = \frac{s_{\hat{y}}^2}{s_y^2} \qquad (6.6)$$

where r^2 is the coefficient of determination, $s_{\hat{y}}^2$ is the variance of the predicted values of the dependent variable, and s_y^2 is the variance of the observed values of the dependent variable.

When there is a perfect fit, the predicted values are the same as the observed values, so that $s_{\hat{y}}^2$ and s_y^2 are identical, giving an r^2 of 1.0. If the fit of the regression is less than perfect, there are differences between predicted and observed values, and $s_{\hat{y}}^2$ is less than s_y^2, giving an r^2 of less than 1.0.

In this case the coefficient of determination can be calculated as:

$$r^2 = \frac{s_{\hat{y}}^2}{s_y^2} = \frac{119.8}{146.2} = 0.819$$

Converting this ratio to a percentage, it can be said that 81.9% of the variance of the dependent variable is accounted for by the regression.

Exercise 6.2

Calculate s_y^2, $s_{\hat{y}}^2$ and r^2 for the regression fitted in Exercise 6.1.

This method of calculating the coefficient of determination has been described in some detail to illustrate the principle behind it. If the product-moment correlation between the two variables has already been calculated, the coefficient of determination can be obtained very simply, by squaring the correlation coefficient. For alders and moisture content the product-moment correlation coefficient has been calculated, in Section 5.1, as 0.905. The coefficient of determination is therefore $0.905^2 = 0.819$.

Tests on the Residuals

The coefficient of determination is a measure of the proportion of the total variation in the data which is accounted for by the regression. A complementary test of goodness of fit involves examining the variation which has **not** been accounted for. This *residual variation* is seen in the scatter of the residuals about the regression line. There is almost certain to be some residual variation, but if the regression has successfully accounted for the variation in the dependent variable, there should be no **systematic** variation in the residuals.

Figure 6.10 shows a number of possible arrangements of residuals. In Figure 6.10a, the residuals are arranged in a way which suggests that a curved line would fit the data better than a straight line. The regression line shown is the best linear fit to the data, but the residuals exhibit a clear systematic variation. From left to right, the residuals decrease in value, and become negative (below the regression line) before gradually increasing and becoming positive again.

In Figure 6.10b the residuals, going from left to right along the regression line, are randomly distributed about the line, in the sense that there is no apparent rule for determining whether any particular residual is positive rather than negative. However, the residuals tend to increase in absolute size from left to right. This would suggest that regression assumption 3 is invalid in this case.

Figure 6.10c illustrates a situation in which there is no systematic variation in the absolute size of the residuals or in the sequence of positive and negative residuals. Examination of the residuals can therefore, reveal violations of the assumptions of regression, as well as whether the regression has accounted for all the systematic variation in the data. Two kinds of test are necessary here:

(1) Tests of whether the sequence of positive and negative residuals is random.

(2) Tests of whether there is any systematic variation in the absolute size of residuals.

One test of each type will be discussed here.

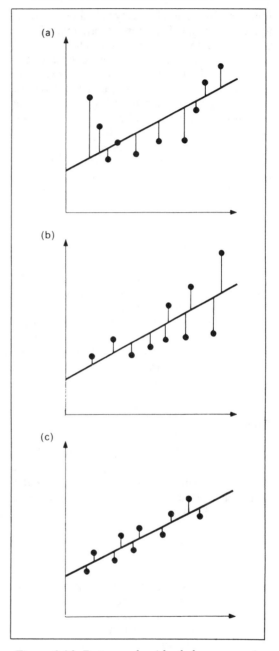

Figure 6.10 Patterns of residuals from regression

A Runs Test for the Order of Residuals

The runs test, described in Section 4.2, is a very simple test of the randomness of a sequence of two alternatives. In the case of regression it can be used to test whether the sequence of positive and negative residuals is random.

For the regression of alders on moisture content, the sequence of residuals, going from left to right along the regression line, is:

$$- + + - - + + + - -$$

The number of runs in this sequence is 5. Where

there are five items of each type in a sequence of ten items, the smallest possible number of runs is two, and the largest is ten. The observed number of runs, is therefore, neither smaller nor larger than might be expected by chance. The null hypothesis, that the sequence is random, can be tested by reference to tables of the critical values of the number of runs. Since the details of this procedure are discussed at length in Section 4.2, only a brief summary of the application of the runs test to this example is given below:

H_0: The sequence of positive and negative residuals is random.

H_1: The sequence of positive and negative residuals is not random.

Significance level: 0.05

Critical values (from Appendix C2):

lower critical value = 2
upper critical value = 10

Observed number of runs: r = 5

Test: Since r is greater than lower critical value and less than upper critical value, the null hypothesis must be accepted.

Conclusion: The sequence of positive and negative residuals is random.

Exercise 6.3

Using the data from Exercises 6.1 and 6.2, calculate the residuals from the regression of y on x. Apply the runs test to find out whether the sequence of positive and negative residuals is random at the 0.05 level.

Autocorrelation

If there is no systematic variation in the residuals, then there should be no correlation between the absolute values of successive residuals. This form of correlation is known as *serial correlation*, or *autocorrelation*. The calculation of the autocorrelation in the residuals from the regression of alders on moisture content is given in Table 6.3. Each pair of values used in this correlation consists of two adjacent residuals. The first residual is paired with the second, the second with the third and so on up to the ninth paired with the tenth. Given ten residuals, nine pairings are produced.

The value of r, in this case 0.351, is a measure of the extent to which the residuals tend either to increase or decrease in absolute value along the regression line. A value close to 1.0 or −1.0 suggests the presence of a relationship between successive residuals. A complete absence of this relationship would give a value of r of 0.0.

In the case of the regression of alders on moisture content the runs test and the test for autocorrelation both suggest that the residuals are indeed random. From this it can be concluded that the linear regression equation is the best possible equation for predicting moisture content from the number of alders.

Exercise 6.4

Using the data from the previous exercises, calculate the degree of autocorrelation in the residuals from the regression of y on x.

Confidence Limits

In Section 3.2 some methods for assessing the reliability of sample estimates were discussed. It was pointed out that, for example, a sample mean is an estimate of the population mean. However, on the basis of sample evidence it is possible to establish limits within which the population value being estimated can be expected to lie. A similar procedure can be applied to predictions made from a regression equation. For the alders/moisture content data, the regression equation (6.5) is a sample estimate of the population regression equation. When the regression equation is used in order to predict the expected value of the dependent variable for any given value of the independent variable, a confidence interval for that expected value can be calculated.

Figure 6.11 shows two such confidence intervals fitted to the regression of alders on moisture content. The 95% confidence interval is the shaded area between the innermost pair of curved lines. For any particular value of x (x_0) the expected value of y ($\hat{y}$) can be predicted. There is a probability of 0.95 (95%) that the value of $\hat{y}$ that would have been obtained had the entire population been available for analysis

Table 6.3 Calculation of Autocorrelation for the Regression of Moisture Content on Alders

First nine a	Last nine b	a^2	b^2	ab
10.00	6.66	100.00	44.36	66.60
6.66	2.22	44.36	4.93	14.79
2.22	0.91	4.93	0.83	2.02
0.91	5.26	0.83	27.67	4.79
5.26	7.04	27.67	49.56	37.03
7.04	3.52	49.56	12.39	24.78
3.52	2.43	12.39	5.90	8.55
2.43	4.00	5.90	16.00	9.72
4.00	1.70	16.00	2.89	6.80

Heading spanned across a^2, b^2 and ab: Absolute values of residuals (i.e. minus signs ignored)

$$\sum a = 42.04 \quad \sum b = 33.74 \quad \sum a^2 = 261.64 \quad \sum b^2 = 164.53 \quad \sum ab = 175.08$$

$$n = 9$$

$$\bar{a} = \frac{42.04}{9} = 4.671 \quad \bar{b} = \frac{33.74}{9} = 3.749$$

$$s_a = \sqrt{\frac{\sum a^2}{n} - \bar{a}^2} = \sqrt{\frac{261.64}{9} - 4.671^2} = \sqrt{29.071 - 21.818} = \sqrt{7.253} = 2.693$$

$$s_b = \sqrt{\frac{\sum b^2}{n} - \bar{b}^2} = \sqrt{\frac{164.53}{9} - 3.749^2} = \sqrt{18.281 - 14.055} = \sqrt{4.226} = 2.056$$

$$r = \frac{\dfrac{\sum ab}{n} - \bar{a}\bar{b}}{s_a s_b} = \frac{\dfrac{175.08}{9} - (4.671 \times 3.749)}{2.693 \times 2.056} = \frac{19.453 - 17.512}{5.537} = 0.351$$

lies within the calculated 95% confidence interval. The difference between the value of $\hat{y}$ computed from the sample and its actual population value is attributable to sampling fluctuations, measurement error, and deviation from the assumptions of regression. It is possible to calculate an interval for any desired level of confidence. Commonly, 95% or 99% intervals are used, as shown in Figure 6.11, but there is no reason why other intervals should not be used.

The confidence interval is given by:

$$t\sqrt{\frac{\sum e^2}{n-2}\left[\frac{1}{n} + \frac{(x_0 - \bar{x})^2}{\sum x^2 - n\bar{x}^2}\right]} \qquad (6.7)$$

Where $\sum e^2$ is the sum of the squares of the residuals from the regression, $\bar{x}$ is the mean of the values of the independent variable, x is the independent variable, x_0 is the particular value of the independent variable for which the expected value of the dependent variable ($\hat{y}$) is being predicted, n is the number of pairs of measurements, and t is a particular value taken from tables of the Student's t distribution (Appendix C4). The choice of a value for t is described in the example which follows. This equation may look rather complex, but in fact almost all the elements of it will already have been calculated in most cases.

For the alders/moisture regression the confidence intervals for $\hat{y}$ will be calculated for five selected values of x_0 of 30, 40, 50, 60, and 70.

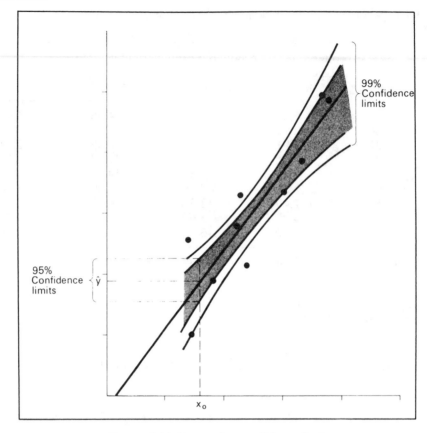

Figure 6.11 Regression confidence limits

The calculation of the various elements of Equation 6.7 is shown in Table 6.4.

All that remains is to substitute the appropriate values of t and x_0 into the equation. The appropriate values of t are:

(1) For the 95% confidence interval, t = the two-tailed critical value of t with n–2 degrees of freedom at the 0.05 significance level.

(2) For the 99% confidence interval, t = the two-tailed critical value of t with n − 2 degrees of freedom at the 0.01 significance level.

For the example data the value of t for the 95% confidence limits is therefore 2.31 (n = 10 gives 8 degrees of freedom). The 95% confidence interval can now be calculated for each value of x_0:

$x_0 = 30$ confidence interval

$$= 2.31 \sqrt{33.07\left[0.1 + \frac{(30-50)^2}{2818}\right]}$$
$$= 6.53$$

$x_0 = 40$ confidence interval

$$= 2.31 \sqrt{33.07\left[0.1 + \frac{(40-50)^2}{2818}\right]}$$
$$= 4.89$$

$x_0 = 50$ confidence interval

$$= 2.31 \sqrt{33.07\left[0.1 + \frac{(50-50)^2}{2818}\right]}$$
$$= 4.20$$

$x_0 = 60$ confidence interval

$$= 2.31 \sqrt{33.07\left[0.1 + \frac{(60-50)^2}{2818}\right]}$$
$$= 4.89$$

Table 6.4 Calculation of Confidence Intervals for Regression of Moisture Content on Alders

Residuals		Values of independent variable	
e	e^2	x	x^2
10.00	100.00	27	729
−6.66	44.36	28	784
−2.22	4.93	35	1225
0.91	0.83	44	1936
5.26	27.67	45	2025
−7.04	49.56	47	2209
−3.52	12.39	60	3600
−2.43	5.90	66	4356
4.00	16.00	73	5329
1.70	2.89	75	5625

$$\sum e^2 = 264.53 \qquad \sum x = 500 \qquad \sum x^2 = 27818$$
$$n = 10 \qquad\qquad \bar{x} = 50$$

$$\text{Confidence interval} = t\sqrt{\frac{\sum e^2}{n-2}\left[\frac{1}{n}+\frac{(x_0-\bar{x})^2}{\sum x^2 - n\bar{x}^2}\right]}$$

$$= t\sqrt{\frac{264.53}{8}\left[\frac{1}{10}+\frac{(x_0-50)^2}{27818-(10\times 50^2)}\right]} = t\sqrt{33.07\left[0.1+\frac{(x_0-50)^2}{2818}\right]}$$

$x_0 = 70$ confidence interval

$$= 2.31\sqrt{33.07\left[0.1+\frac{(70-50)^2}{2818}\right]}$$

$$= 6.53$$

The confidence interval can be calculated for any x_0 value, though it would not be advisable to choose an x_0 which lies outside the observed range of the independent variable. For example, when the value of the independent variable (number of alders) is 40, the expected value of the dependent variable (moisture content) can be predicted from the regression Equation (6.5) as:·

$$\hat{y} = -2.6+0.652x = -2.6+(0.652\times 40)$$
$$= 23.48$$

However, this value is only an estimate based on sample data. Applying the 95% confidence interval for $x_0 = 40$ shows that there is a 95% (0.95) probability that the population value will be 23.48±4.89. In other words, the population value can be expected to lie somewhere between 18.59 (23.48−4.89) and 28.37 (23.48+4.89) with 95% confidence. Similarly, when the number of alders is 70, the expected value for moisture content from the regression equation is:

$$-2.6+(0.652\times 70) = 43.04$$

With 95% confidence the population value can be expected to lie somewhere between 36.51 (43.04−6.53) and 49.57 (43.04+6.53).

Exercise 6.5

(a) Within what limits can the population value for moisture content be expected to lie with 95% confidence when the number of alders is 65?

(b) What value of t is appropriate for calculating the 99% confidence interval for the regression of moisture content on alders?

(c) Within what limits can the population value for moisture content be expected to lie with 99% confidence when the number of alders is (i) 50, (ii) 60, and (iii) 65?

By calculating confidence intervals for $\hat{y}$ for selected values of the independent variable it is possible to build up confidence bands like the ones illustrated in Figure 6.11. It can be seen that these bands are narrowest at a point on the regression line corresponding to the mean of the observed values of the independent variable. Predictions made from a regression equation are therefore, more reliable for values of the independent variable close to their mean. For values further away from this mean predictions become progressively less reliable.

Exercise 6.6

Calculate the 95% confidence interval for $\hat{y}$ for each observed value of x in the regression calculated for exercise 6.1 (data from Table 5.2).

6.2 Nonlinear Regression

Figure 6.12 represents a set of data obtained from a sample survey of visitors to a National Park. The number of visitors from each of twenty five-mile zones around the park (on the vertical axis) is plotted against distance (on the horizontal axis). It can be seen that there is some kind of relationship between number of visitors and distance from the park. With increasing distance there is a decline in the number of visitors attracted. However, this decline is not uniform; it could not be represented by a straight line on a graph. This is an example of the sort of nonlinear relationship that is quite common in human geography.

Figure 6.13 is a graph based on the same set of data. This time, instead of plotting the number of visitors against distance, the logarithm of the number of visitors is plotted against the logarithm of the distance. By transforming the two variables, number of visitors and distance, from an arithmetic to a logarithmic scale, a nonlinear relationship has been changed into a linear relationship. This procedure is not quite as neat as it appears at first sight. There remains, for example, the problem

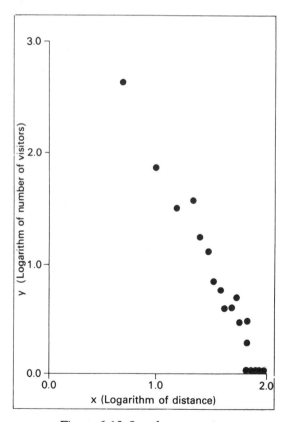

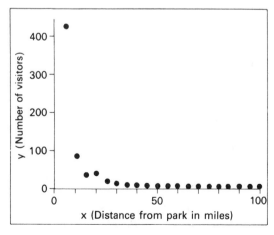

Figure 6.12 Nonlinear regression

Figure 6.13 Log-log regression

Table 6.5 Regression of Number of Visitors on Distance Travelled to National Park

Original values distance x	visitors y	Transformed values x′	y′	x′²	x′y′
5	424	0.6990	2.6274	0.4886	1.8366
10	77	1.0000	1.8865	1.0000	1.8865
15	32	1.1761	1.5051	1.3832	1.7701
20	36	1.3010	1.5563	1.6926	2.0247
25	17	1.3979	1.2304	1.9541	1.7200
30	13	1.4771	1.1139	2.1818	1.6453
35	7	1.5441	0.8451	2.3842	1.3049
40	6	1.6021	0.7782	2.5667	1.2468
45	4	1.6532	0.6021	2.7331	0.9954
50	4	1.6990	0.6021	2.8866	1.0230
55	5	1.7404	0.6990	3.0290	1.2165
60	3	1.7781	0.4771	3.1616	0.8483
65	3	1.8129	0.4771	3.2866	0.8649
40	2	1.8451	0.3010	3.4044	0.5554
75	1	1.8751	0.0000	3.5160	0.0000
80	1	1.9031	0.0000	3.6218	0.0000
85	1	1.9294	0.0000	3.7226	0.0000
90	1	1.9542	0.0000	3.8189	0.0000
95	1	1.9777	0.0000	3.9113	0.0000
100	1	2.0000	0.0000	4.0000	0.0000
		32.3655	14.7013	54.7431	18.9384

$$\sum x' = 32.37 \qquad \sum y' = 14.70 \qquad \sum x'^2 = 54.74 \qquad \sum x'y' = 18.94$$

$$\bar{x}' = \frac{\sum x'}{n} = \frac{32.37}{20} = 1.618 \qquad \bar{y}' = \frac{\sum y'}{n} = \frac{14.70}{20} = 0.735$$

note: $x' = \log_{10}x$, $y' = \log_{10}y$

of rewriting the linear relationship between the transformed variables in terms of the original, 'raw' variables. In the present case there is a linear relationship between the log of the number of visitors and the log of distance. To be of practical use to the geographer, this needs to be translated into some form of relationship between of visitors and distance.

Applying to the technique of linear regression to the transformed data will yield a regression of the form:

$$\log_{10}\hat{y} = a + b\log_{10}x \qquad (6.8)$$

By taking the antilogarithm of both sides of this equation, it is possible to rewrite it in terms of x and y instead of $\log_{10}x$ and $\log_{10}y$:

$$\hat{y} = \text{antilog}_{10}ax^b \qquad (6.9)$$

Equation 6.8 could be used to predict the value of the logarithm of the number of visitors, given the logarithm of the distance. Equation 6.9 is rather more useful, as it enables distance to be used to predict number of visitors directly.

Table 6.5 contains the data from which Figure 6.12 was drawn. It also gives the logarithms of all the values of the two variables. It is these logarithms to which the linear regression technique is to be applied. Effectively this means fitting a least-squares regression line to the graph shown in Figure 6.13. The regression coefficients can be calculated using the equations given previously:

$$b = \frac{\sum x'y' - n\bar{x}'\bar{y}'}{\sum x'^2 - n\bar{x}^2} \qquad a = \bar{y}' - b\bar{x}'$$

The prime (') has been added to x and y to denote that these are variables which have been transformed, in this case by taking logarithms. In other words, x' (pronounced x prime) means $\log_{10}x$ and y' means $\log_{10}y$, in this example.

The various elements of the two equations given above have been calculated in Table 6.5. these can be substituted into the equations in order to give the regression coefficients:

$$b = \frac{18.938 - (20 \times 1.618 \times 0.735)}{54.743 - (20 \times 1.618^2)} = -2.06$$

$$a = 0.735 - (-2.06 \times 1.618) = 4.07$$

The regression equation can now be written out:

$$\hat{y}' = 4.07 - 2.06x' \qquad (6.10)$$

or since x' and y' are $\log_{10}x$ and $\log_{10}y$ respectively:

$$\log_{10}\hat{y} = 4.07 - 2.06 \log_{10}x \qquad (6.11)$$

Equation 6.11 could now be used to predict the expected value of $\log_{10}y$ for any given value of $\log_{10}x$. By taking the antilogarithms of both sides of the equation, however, it is possible to obtain an equation for predicting y from x:

$$\hat{y} = \text{antilog}_{10}ax^b$$

which in this case means:

$$\hat{y} = \text{antilog}_{10}4.07x^{-2.06} \qquad (6.12)$$

From standard mathematical tables $\text{antilog}_{10}4.07$ is found to be 11750, so that Equation 6.12 can be further simplified to:

$$\hat{y} = 11750x^{-2.06} \quad \text{or} \quad \hat{y} = \frac{11750}{x^{2.06}}$$

$$(6.13)$$

This relationship can be put into words as follows: the expected number of visitors is equal to 11750 divided by the distance to the power of 2.06. For example, approximately 102 visitors

could be expected to come from a distance of 10 miles:

$$\hat{y} = \frac{11750}{x^{2.06}} = \frac{11750}{10^{2.06}} = \frac{11750}{114.81} = 102.34$$

This compares with the actual number of visitors coming from the 10 mile zone, which is 77.

Exercise 6.7

(a) How many visitors would you expect to come from the 50 mile zone, according to the regression equation (6.13)?

(b) How many from the 75 mile zone?

Inferential Aspects

When linear regression is used in conjunction with transformed variables to describe what is originally a nonlinear relationship, the five major assumptions of the technique (discussed in Section 6.1) still apply. However, it is the **transformed** data, not the 'raw' values which must satisfy these assumptions. The assumptions can therefore, be rewritten to fit the present situation:

(1) The independent variable consists of a set of fixed values, not a set of sample measurements, or, alternatively, the values of the independent variable have been measured without error.

(2) The values of the transformed dependent variable y' are normally distributed.

(3) The variance of the transformed values of the dependent variable is constant for all values of the independent variable.

(4) The values of the residuals from the regression between the transformed variables have a normal distribution.

(5) The values of the residuals from the regression between the transformed variables are independent of each other.

As far as the National Park example is concerned, the first assumption is satisfied, since the distance zones were fixed values. The implication of the second assumption is rather more difficult to appreciate. Suppose the sample survey were to be repeated a large number of times. A record could be kept of the logarithm of the number of visitors from each zone as revealed by each survey. For each zone a frequency distribution could then be constructed to show how often any particular transformed value of the number of visitors occurred. The second assumption means that each of these frequency distributions of transformed values should be found to be normal. What the second assumption does **not** imply is that the frequency distribution of the raw values of the dependent variable for any one zone is normal. Following on from this, the third assumption means that the variance of each of these frequency distributions (one for each zone) should be the same.

The fourth and fifth assumptions both concern the pattern of residuals from the regression between the transformed variables. In this case they concern the pattern of residuals from the regression of the logarithm of the number of visitors on the logarithm of distance. These residuals are calculated in Table 6.6. As with a normal regression, these residuals can be tested for randomness and serial correlation using the techniques described in Section 6.1. These tests provide some measure of whether the particular transformation chosen, in this case a logarithmic transformation, has successfully enabled the relationship to be summarized by a regression equation.

Exercise 6.8

(a) Use the runs test to find out whether the sequence of positive and negative residuals from the regression of log number of visitors on log distance is random at the 0.05 level. The residuals are given in Table 6.6.

(b) Calculate the degree of serial correlation between these residuals.

Table 6.6 Observed, Predicted and Residual Values from Regression of Number of Visitors on Distance

Observed y'	Predicted $\hat{y}'$	Residual e	y'^2	$\hat{y}'^2$	e^2
2.6274	2.6031	−0.0027	6.9032	6.9137	0.0000
1.8865	2.0100	−0.1235	3.5589	4.0401	0.0153
1.5051	1.6472	−0.1421	2.2653	2.7133	0.0202
1.5563	1.3899	0.1664	2.4221	1.9318	0.0277
1.2304	1.1903	0.0401	1.5139	1.4168	0.0016
1.1139	1.0272	0.0867	1.2408	1.0551	0.0075
0.8451	0.8892	−0.0441	0.7142	0.7907	0.0019
0.7782	0.7697	0.0085	0.6056	0.5924	0.0001
0.6021	0.6644	−0.0623	0.3625	0.4414	0.0039
0.6021	0.5701	0.0320	0.3625	0.3250	0.0010
0.6990	0.4848	0.2142	0.4886	0.2350	0.0459
0.4771	0.4071	0.0700	0.2276	0.1657	0.0049
0.4771	0.3354	0.1417	0.2276	0.1125	0.0201
0.3010	0.2691	0.0319	0.0906	0.0724	0.0010
0.0000	0.2073	−0.2073	0.0000	0.0430	0.0430
0.0000	0.1496	−0.1496	0.0000	0.0224	0.0224
0.0000	0.0954	−0.0954	0.0000	0.0091	0,0091
0.0000	0.0443	−0.0443	0.0000	0.0020	0.0020
0.0000	−0.0040	0.0040	0.0000	0.0000	0.0000
0.0000	−0.0499	0.0499	0.0000	0.0025	0.0025
			20.9834	20.8886	0.2301

$\sum y'^2 = 20.98$ $\sum \hat{y}'^2 = 20.89$ $\sum e^2 = 0.23$

(c) Calculate the coefficient of determination for this regression. Most of the necessary figures are given in Tables 6.5 and 6.6.

Confidence Limits

When dealing with a regression between transformed variables, confidence limits can be calculated in the usual way, provided the transformed values rather than the raw values of the variables are used. The equation for the confidence interval thus becomes:

$$t\sqrt{\frac{\sum e^2}{n-2}\left[\frac{1}{n}+\frac{(x_0'-\bar{x}')^2}{\sum x'^2-n\bar{x}'^2}\right]}$$

This is similar to the equation for the confidence interval given previously (Equation 6.7), except that x' is substituted for x to indicate that transformed values are being used. Similarly the values of the residuals are based on the transformed values:

$$e = y' - \hat{y}' \qquad (6.15)$$

To calculate the 95% confidence interval for the regression of log number of visitors on log distance:

(1) The appropriate value of t for use in Equation 6.4 is the two-tailed critical value of t with 18 degrees of freedom at the 0.05 significance level. From tables of critical values of t (Appendix C4) t is found to be 2.10.

(2) The various elements of Equation 6.4 have already been calculated in Tables 6.5 and 6.6. They can be substituted into Equation 6.14:

confidence interval (95%)

$$= 2.10\sqrt{\frac{0.230}{20-2}\left[\frac{1}{20}+\frac{(x_0'-1.618)^2}{54.743-(20\times 1.618^2)}\right]}$$

$$= 2.10\sqrt{0.013\left[0.050+\frac{(x_0'-1.618)^2}{2.385}\right]}$$

(3) The required values of x_0' can now be substituted into the equation given above. For

example, if $x_0' = 0.6990$ (i.e. $x = 5$), the 95% confidence interval is given by:

$$2.10\sqrt{0.013\left[0.050+\frac{(0.699-1.618)^2}{2.385}\right]}$$

$$= 2.10\sqrt{0.013\left[0.050+\frac{-0.919^2}{2.835}\right]}$$

$$= 2.10\sqrt{0.013\left[0.050+\frac{0.845}{2.835}\right]}$$

$$= 2.10\sqrt{0.013(0.050+0.354)}$$

$$= 2.10\sqrt{0.013\times 0.404} = 2.10\sqrt{0.005}$$

$$= 2.10\times 0.071 = 0.149$$

In other words, the population value of y' can be expected to lie within ± 0.149 of the predicted value with 95% confidence. Since the predicted value has already been calculated as 2.6301 (Table 6.6), the population value can be expected, with 95% confidence, to lie between 2.4811 (2.6301 − 0.149) and 2.7791 (2.6301 + 0.149).

(4) The confidence interval can be 'translated' to relate to the raw values of the dependent variable, in this case by taking antilogarithms:

antilog$_{10}$2.4811 = 302.8 (lower 95% limit)

antilog$_{10}$2.7791 = 601.3 (upper 95% limit)

Figure 6.14a shows the 95% confidence limits fitted to the regression of $\log_{10}$ number of visitors on $\log_{10}$ distance. In Figure 6.14b the confidence limits have been 'translated' to fit the original graph of visitors against distance. It can be seen that the 'translated' confidence limits have a very strange shape. This is a result of the fact that the limits have been fitted to the transformed data values, not to the raw values. It is important to realize that the whole regression in this context relates to the transformed values first. It is the transformed values that must satisfy the assumptions of regression, not the raw values. Similarly the confidence limits apply first to the transformed values. When

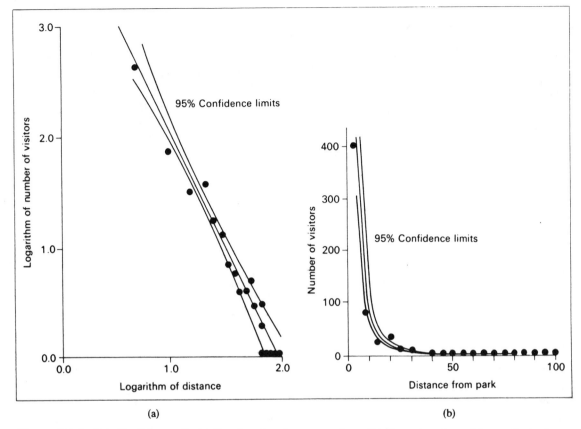

Figure 6.14   (a) Confidence limits fitted to log–log regression. (b) Translated confidence limits from nonlinear regression

translated to give limits in terms of the original scales of measurement, the results may be difficult to appreciate, and may indeed defy common sense.

Numerous other transformations can be applied to data to make them conform to a straight line relationship, and thereby make them suitable for linear regression. It is possible to replace all the values by their square roots or their squares. It is even possible to apply a different transformation to one variable from

that applied to the other. It is not difficult to see that in doing this there is a very great danger that the geographer will entirely lose sight of his research aim. It is much easier to overlook the shortcomings of a set of data, and to forget about the measurement errors involved in their collection, if the data no longer refer to numbers of people or distances, but instead to logarithms or square roots. Perhaps the soundest advice is to treat nonlinear regression based on data transformations with extreme caution.

6.3 Computer Program

Program II — Simple Linear Regression

The program performs a linear regression of a dependent variable y on an independent variable x. It calculates predicted values of y, residuals, and confidence limits.

Program Listing

```
10 REM - SIMPLE LINEAR REGRESSION
20 PRINT"---------------------------"
30 PRINT"SIMPLE LINEAR REGRESSION"
40 PRINT"---------------------------"
50 PRINT
60 INPUT"Number of values of each variable",NX
70 N=NX+NX
80 DIM X(N)
90 PRINT
100 PRINT"Now type in values of X:"
110 PRINT"(Mistakes can be corrected later)"
120 FOR I=1 TO NX
130 PRINT I;" ";
140 INPUT X(I)
150 NEXT I
160 PRINT
170 PRINT"Now type in values of Y:"
180 PRINT"(Mistakes can be corrected later)"
190 FOR I=NX+1 TO N
200 PRINT I-NX;" ";
210 INPUT X(I)
220 NEXT I
230 N1=1
240 N2=NX
250 V$="X"
260 GOSUB 1100
270 N1=NX+1
280 N2=N
290 V$="Y"
300 GOSUB 1100
310 FOR I=1 TO NX
320 XM=XM+X(I)
330 YM=YM+X(I+NX)
340 SX=SX+X(I)^2
350 SY=SY+X(I+NX)^2
360 XY=XY+X(I)*X(I+NX)
370 NEXT I
380 XM=XM/NX
390 YM=YM/NX
400 B=(XY-NX*XM*YM)/(SX-NX*XM^2)
410 A=YM-B*XM
420 X2=SX-NX*XM^2
430 SX=SQR(SX/NX-XM^2)
440 SY=SQR(SY/NX-YM^2)
450 R=(XY/NX-XM*YM)/(SX*SY)
460 PRINT
470 PRINT"---------------------------"
480 PRINT"SIMPLE LINEAR REGRESSION"
490 PRINT"---------------------------"
500 PRINT
510 PRINT"Mean of X = ";XM
520 PRINT"Standard deviation of X = ";SX
530 PRINT"Mean of Y = ";YM
540 PRINT"Standard deviation of Y = ";SY
550 PRINT
560 PRINT"r = ";R
570 DF=NX-2
580 PRINT"with ";DF;" degrees of freedom"
590 PRINT
600 PRINT"Regression of Y on X:"
610 A$=" + "
620 IF B<0 A$=" "
630 PRINT"Y = ";A;A$;B;" X"
640 PRINT"Coefficient of determination = ";R^2
650 PRINT
660 PRINT"Do you want predicted and ";
670 INPUT"residual values (Y/N)";Q$
680 IF Q$="N" GOTO 810
690 IF Q$<>"Y" GOTO 660
700 PRINT
710 PRINT"        X         Y    Predicted  Residual"
720 PRINT
730 E=0
740 FOR I=1 TO NX
750 YP=A+B*X(I)
760 R=X(I+NX)-YP
770 E=E+R^2
780 PRINT X(I),X(I+NX),YP,R
790 NEXT I
800 E=E/DF
810 PRINT
820 PRINT"Do you want confidence limits";
830 INPUT" (Y/N)";Q$
840 IF Q$="N" GOTO 1090
850 IF Q$<>"Y" GOTO 820
860 PRINT
870 PRINT"Type in critical values of Student's t"
880 PRINT"with ";DF;" degrees of freedom:"
890 PRINT
900 INPUT"At 0.05 level (two-tailed): t = ";T1
910 INPUT"At 0.01 level (two-tailed): t = ";T2
920 N1=1/NX
930 PRINT
940 PRINT"Now type in X value for each confidence inter
950 PRINT
960 INPUT"X = ";XI
970 PRINT
980 YP=A+B*XI
990 C1=T1*SQR(E*(N1+(XI-XM)^2/X2))
1000 C2=T2*SQR(E*(N1+(XI-XM)^2/X2))
1010 PRINT"95% confidence limits:"
1020 PRINT"Y = ";YP;" + or - ";C1
1030 PRINT"99% confidence limits:"
1040 PRINT"Y = ";YP;" + or - ";C2
1050 PRINT
1060 INPUT"More (Y/N)";Q$
1070 IF Q$="Y" GOTO 950
1080 IF Q$<>"N" GOTO 1060
1090 STOP
1100 REM - DATA CHECKING ROUTINE
1110 PRINT
1120 PRINT"Do you want to check/correct ";V$;
1130 INPUT" values (Y/N)";Q$
1140 IF Q$="N" GOTO 1250
1150 IF Q$<>"Y" GOTO 1120
1160 K=0
1170 PRINT"Press RETURN if value is correct";
1180 PRINT" - otherwise type correct value"
1190 FOR I=N1 TO N2
1200 K=K+1
1210 PRINT K;" ";X(I);
1220 INPUT A$
1230 IF A$<>"" X(I)=VAL(A$)
1240 NEXT I
1250 RETURN
```

Example Output

```
------------------------
SIMPLE LINEAR REGRESSION
------------------------

Number of values of each variable?10

Now type in values of X:
(Mistakes can be corrected later)
        1 ?47
        2 ?35
        3 ?28
        4 ?27
        5 ?44
        6 ?66
        7 ?60
        8 ?75
        9 ?73
       10 ?45

Now type in values of Y:
(Mistakes can be corrected later)
        1 ?21
        2 ?18
        3 ?9
        4 ?25
        5 ?27
        6 ?38
        7 ?33
        8 ?48
        9 ?49
       10 ?32

Do you want to check/correct X values (Y/N)?N

Do you want to check/correct Y values (Y/N)?N

------------------------
SIMPLE LINEAR REGRESSION
------------------------

Mean of X = 50
Standard deviation of X = 16.7869
Mean of Y = 30
Standard deviation of Y = 12.09132

r = 0.9050342
with 8 degrees of freedom

Regression of Y on X:
Y = -2.594038 + 0.6518808 X
Coefficient of determination = 0.8190868

Do you want predicted and residual values (Y/N)?Y

        X      Y  Predicted   Residual

        47     21  28.04436   -7.044358
        35     18  20.22179   -2.221789
        28      9  15.65862   -6.658623
        27     25  15.00674    9.993258
        44     27  26.08872    0.9112846
        66     38  40.43009   -2.430092
        60     33  36.51881   -3.518808
        75     48  46.29702    1.702981
        73     49  44.99326    4.006742
        45     32  26.7406     5.259404
```

```
Do you want confidence limits (Y/N)?Y

Type in critical values of Student's t
with 8 degrees of freedom:

At 0.05 level (two-tailed): t = ?2.31
At 0.01 level (two-tailed): t = ?3.35

Now type in X value for each confidence interval:

X = ?30

95% confidence limits:
Y = 16.96238 + or - 6.533317
99% confidence limits:
Y = 16.96238 + or - 9.474724

More (Y/N)?Y

X = ?40

95% confidence limits:
Y = 23.48119 + or - 4.889036
99% confidence limits:
Y = 23.48119 + or - 7.09016

More (Y/N)?Y

X = ?50

95% confidence limits:
Y = 30 + or - 4.200256
99% confidence limits:
Y = 30 + or - 6.09128

More (Y/N)?N

STOP at line 1090
```

The data used are for the relationship between moisture content and number of alders (Table 6.1).

7 Spatial Statistics

Although there may be disagreement in detail about the nature and purpose of geography, it can reasonably be argued that one of the central themes of geographical enquiry is the distribution of phenomena on the Earth's surface. In view of the enthusiasm with which geographers embraced the 'quantitative revolution', it is perhaps surprising that remarkably little progress has been made in the development and application of spatial statistics. Although a wide variety of statistical techniques has been applied by geographers to data which have been collected on an areal basis, only a handful of techniques are in common use for analysing the spatial distribution of these data.

There are several possible explanations for this state of affairs. First there is the lack of pre-existing theory and methods of statistical analysis applicable to spatial distributions. Statisticians themselves have not shown a great deal of interest in spatial statistics in the past, and geographers are in a minority in demanding them now. Secondly many of the techniques which do exist are, at least at first sight, difficult to understand and tedious to use. It is certainly true that some of the more advanced spatial techniques can not be applied to other than trivial problems without the aid of a computer.

Despite these reservations there are several simple spatial techniques which can be applied using manual methods of calculation, and which are intuitively easy to understand. It may well be that the present underuse of spatial techniques is merely due to lack of publicity. Various types of phenomena can be studied using these techniques: points, lines, and areas, and a number of different characteristics can be measured. These include central tendency, dispersion, shape, pattern and spatial relationships.

7.1 Central Tendency in Point Patterns

In Chapter 2 various measures of *central tendency* were applied to sets of non-spatial data. Each of these measures was described as giving some indication of the 'average' value in a set of data, or the 'centre' of a frequency distribution. When dealing with spatial distributions the concept of a 'centre' is intuitively reasonable, but there are several ways in which the position of such a centre can be calculated, each of which will give a different result. It is important to realize that there is no one 'correct' answer to the problem of finding the centre of a spatial distribution. Each measure has a different interpretation and the choice should be determined by the nature of the problem.

The Mean Centre

The *mean centre* is the simplest measure of the centre of a spatial distribution. It is analogous to the mean of a set of data, and is calculated in a very similar way.

Figure 7.1 shows a spatial distribution of points. It could be the distribution of towns or drumlins, or of any other geographical phenomenon. As a first step in calculating the

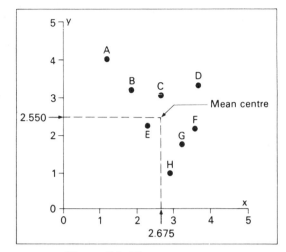

Figure 7.1 The mean centre

mean centre it is necessary to devise some way of quantifying the locations of the points. This can be done by calculating the co-ordinates of each point according to an arbitrary system. In practice the geographer will often already know the map co-ordinates of the points. For example, using Ordnance Survey maps it is a relatively simple matter to measure the position of any point in kilometres east and north of the bottom left-hand corner of the map.

For the calculation of most spatial statistics the position of points needs to be measured in relation to some such co-ordinate system. The orientation of the co-ordinate grid is quite arbitrary, however. Geographers are used to measuring location in terms of eastings and northings, but there is no reason why they should not use, say, southeastings and northeastings. Similarly the origin of the grid, the point from which the co-ordinates are measured, is arbitrary. The origin of the British National Grid, for example, is a point some 150 kilometres west of the Lizard peninsula. The only prerequisites of a co-ordinate system which is to be used in the calculation of spatial statistics are:

(1) the co-ordinate axes must be at right angles to each other, in other words they must be *orthogonal* axes

(2) measurements along the two axes must be made in the same units.

In Figure 7.1 an arbitrary co-ordinate system has been superimposed, with its origin at the

bottom left-hand corner. For simplicity the horizontal axis, measuring eastings, has been labelled x, and the vertical axis, measuring northings, has been labelled y. This convention for labelling co-ordinate axes will be followed throughout the chapter. The axes have been marked off in arbitrary distance units. Table 7.1 gives the co-ordinates for all points marked in Figure 7.1.

The mean centre can now be found simply by calculating the mean of the x co-ordinates (eastings) and the mean of the y co-ordinates (northings). These two mean co-ordinates mark the location of the mean centre. The equation for the mean centre is thus:

$$\bar{x} = \frac{\sum x}{n}, \qquad \bar{y} = \frac{\sum y}{n} \qquad (7.1)$$

where x and y are the co-ordinates of the points, $\bar{x}$ and $\bar{y}$ are the means of the x and y co-ordinates respectively, and n is the number of points. The calculation of the mean centre for Figure 7.1 is given in Table 7.1, and its position is marked in Figure 7.1.

Table 7.1 Finding the Mean Centre

Point	Co-ordinates	
	x	y
A	1.2	4.0
B	1.8	3.2
C	2.7	3.0
D	3.7	3.2
E	2.3	2.2
F	3.6	2.1
G	3.2	1.7
H	2.9	1.0

n = 8 $\sum x = 21.4$ $\sum y = 20.4$

$$\bar{x} = \frac{21.4}{8} = 2.675 \quad \bar{y} = \frac{20.4}{8} = 2.550$$

Co-ordinates of mean centre = 2.675, 2.550

In Section 2.1 the mean was described as the centre of gravity of a set of data. Using the same analogy the mean centre can be thought of as the centre of gravity of a spatial distribution of points. If Figure 7.1 is thought of as a square of weightless material, and each point is taken to be a small coin, then the mean centre is the point on which the square would balance. This is illustrated in Figure 7.2.

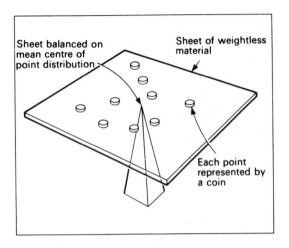

Figure 7.2 *The mean centre as a centre of gravity*

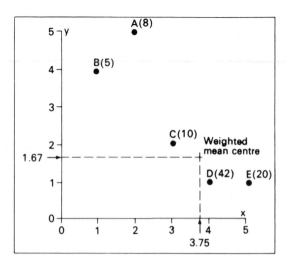

Figure 7.3 *The weighted mean centre*

The Weighted Mean Centre

In the calculation of the mean centre each point is given equal weight. In some situations, however, it may be clear that each point should not be given equal weight when calculating the centre of the distribution. Figure 7.3 shows the distribution of a number of car factories. The

mean centre of this point pattern could easily be calculated in the manner described above. This would give the mean centre of car **factories**. If the number of cars produced by each of the factories were known, it would also be possible to calculate the mean centre of car **production**. Each factory would be given a weight corresponding to the number of cars it produced, so that a factory which produced twice as many cars as another would have twice as much influence on the final location of the mean

Table 7.2 *Finding the Weighted Mean Centre*

Point	Co-ordinates x	y	Weight w	Weighted co-ordinates xw	yw
A	2	5	8	16	40
B	1	4	5	5	20
C	3	2	10	30	20
D	4	1	42	168	42
E	5	1	20	100	20
			85	319	142

$$\bar{x}_w = \frac{\sum xw}{\sum w} = \frac{319}{85.} = 3.75$$

$$\bar{y}_w = \frac{\sum yw}{\sum w} = \frac{142}{85} = 1.67$$

Co-ordinates of weighted mean centre are 3.75, 1.67

centre. A mean centre calculated in this way is known as a *weighted mean centre*.

In Figure 7.3 the numbers in brackets beside each point are weights proportional to the average annual car production of each factory. The weighted mean centre of this distribution can now be found by multiplying the x and y co-ordinates for each point by the weight assigned to that point. The mean of the weighted x co-ordinates and the mean of the weighted y co-ordinates define the position of the weighted mean centre. The equation for the weighted mean centre is thus:

$$\bar{x}_w = \frac{\sum xw}{\sum w}, \qquad \bar{y}_w = \frac{\sum yw}{\sum w} \qquad (7.2)$$

where x and y are the co-ordinates of the points, w denotes the numerical weight assigned to each point, $\bar{x}_w$ and $\bar{y}_w$ are the weighted means of these co-ordinates, and n is the number of points. The calculation of the weighted mean centre of the distribution of car factories shown in Figure 7.3 is given in Table 7.2, and the position of the weighted mean centre is marked in Figure 7.3.

The centre of gravity analogy is also appropriate to the weighted mean centre. If the distribution of car factories shown in Figure 7.3 is imagined as a sheet of weightless material on which each factory is represented as a pile of small coins according to the weight assigned to it, then the weighted mean centre is the point on

which the distribution would balance. This is illustrated in Figure 7.4.

To recap, the mean centre treats each point as being of equal importance whereas the weighted mean centre allows each point to be assigned a weight proportional to its importance. The weights might represent amounts of production or numbers of employees in the case of factories, or population in the case of towns. Whatever characteristic the weights represent, it is the centre of gravity of the distribution of this **characteristic** which is given by the weighted mean centre. The mean centre can only give the centre of gravity of the distribution of **points** themselves.

Exercise 7.1

Figure 7.5 is a map of Nottinghamshire, showing the main towns, with the 1971 population total given in brackets beside each one.

(a) Find the mean centre of the distribution of towns in Nottinghamshire.

(b) Find the centre of gravity of the distribution of urban population in Nottinghamshire, in other words the weighted mean centre of the distribution of towns.

The Median Centre

The term *median centre* has unfortunately two entirely different uses. In most English statistical texts, for example Cole and King (1968) and Hammond and McCullagh (1974), it is defined as the intersection of two orthogonal axes, each of which has an equal number of points on either side. This is the definition which will be taken in this book. The alternative definition, which appears in a number of American texts, such as King (1969) and Neft (1966), will be discussed in the next section as 'the centre of minimum travel'.

The *median centre*, following the English usage of the term, is analogous to the median of a set of data. In Section 2.1 the median was described as the value which has half the data set above it and half below. The median centre is located in such a way that it has as many points to the south as to the north of it, and as many to the west as to the east. This is illustrated in

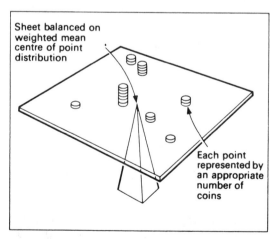

Figure 7.4 The weighted mean centre as a centre of gravity

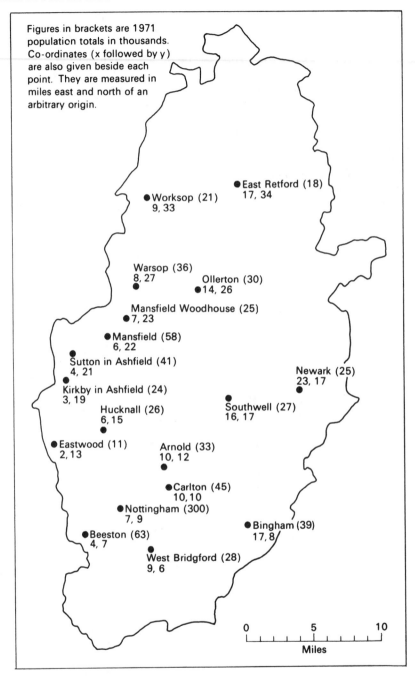

Figures in brackets are 1971 population totals in thousands. Co-ordinates (x followed by y) are also given beside each point. They are measured in miles east and north of an arbitrary origin.

East Retford (18)
17, 34

Worksop (21)
9, 33

Warsop (36)
8, 27

Ollerton (30)
14, 26

Mansfield Woodhouse (25)
7, 23

Mansfield (58)
6, 22

Sutton in Ashfield (41)
4, 21

Newark (25)
23, 17

Kirkby in Ashfield (24)
3, 19

Hucknall (26)
6, 15

Southwell (27)
16, 17

Eastwood (11)
2, 13

Arnold (33)
10, 12

Carlton (45)
10, 10

Nottingham (300)
7, 9

Bingham (39)
17, 8

Beeston (63)
4, 7

West Bridgford (28)
9, 6

0   5   10
Miles

Figure 7.5 Towns in Nottinghamshire

Figure 7.6a, in which the distribution has been divided up by two lines at right angles to each other. The horizontal line has 10 points on either side of it, as does the vertical line.

The advantage of the median centre is that its location can be found very quickly, without resorting to any mathematics other than counting points. The disadvantage is that its location depends on the orientation of the two lines used to divide up the point distribution. This is illustrated in Figure 7.6b, in which the median centre of the same point distribution has been

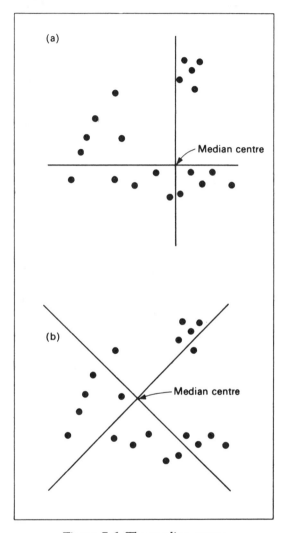

Figure 7.6 The median centre

found using two lines still at right angles to each other, but this time going diagonally across the distribution. Since the location of the median centre can not be uniquely found, its use should be restricted to preliminary geographical investigations, where speed may be more important than accuracy.

The Centre of Minimum Travel

The *centre of minimum travel*, referred to in many American texts as the median centre, is the location from which the sum of the distances to all the points in a distribution is a minimum. This is illustrated in Figure 7.7.

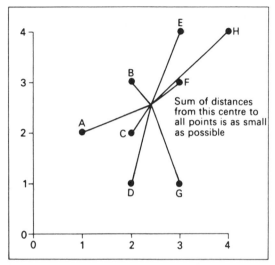

Figure 7.7 Centre of minimum travel

The position of this centre could clearly be found manually by a process of trial and error. By choosing a number of alternative trial locations and calculating the sum of the distances from each trial centre to all the points, the true centre of minimum travel could eventually be found. In most cases the mean centre, median centre and centre of minimum travel are likely to be fairly close to each other. Either of the first two could therefore be used as a starting point in the search for the centre of minimum travel. Although this *iterative* procedure (one which involves a long series of repeated steps) would be extremely time consuming to do manually, it can easily be programmed for a computer.

Exercise 7.2

As a possible class exercise, find the approximate centre of minimum travel for the distribution of towns in Nottinghamshire, shown in Figure 7.5. Each person in the class could try to 'guestimate' the position of the centre of minimum travel and then measure the distance to all the towns. The person with the lowest total distance has come closest to finding the centre of minimum travel.

7.2 Dispersion in Point Patterns

Just as the measures of dispersion discussed in Section 2.2 describe the spread of values around some form of average, so measures of spatial dispersion give information about the areal spread of points around a centre. In this section two techniques for measuring the spread of points around the mean centre will be discussed.

Standard Distance

Standard distance, otherwise known as standard distance deviation or root mean square distance deviation, is the spatial equivalent of standard deviation (Section 2.2). It provides the most concise description of the spread of points around the mean centre. The simplest equation for standard distance is:

$$\text{standard distance} = \sqrt{\frac{\sum d^2}{n}} \qquad (7.3)$$

where d is the distance of each point from the

mean centre and n is the number of points. Having located the mean centre, it is possible to measure all the distances directly from the map, square them, add up all the squares, divide by the number of points and then take the square root. For many map distributions this will be the simplest, and the quickest, way of calculating the standard distance.

An alternative method of calculation involves measuring the distance via Pythagoras' theorem. This method is particularly useful if standard distance is to be calculated by a computer program (see Mather, 1976). Figure 7.8 shows the calculation of the distance between one point and the mean centre of a point pattern. In this approach the squares of the distances, needed for Equation 7.3, are calculated directly from the co-ordinates of the points and of the mean centre. Equation 7.3 can therefore be rewritten as:

standard distance

$$= \sqrt{\frac{\sum (x - \bar{x})^2 + \sum (y - \bar{y})^2}{n}} \qquad (7.4)$$

Using Equation 7.4 would involve rather a lot of calculations, so to reduce the amount of arithmetic it can be modified to:

standard distance

$$= \sqrt{\left(\frac{\sum x^2}{n} - \bar{x}^2\right) + \left(\frac{\sum y^2}{n} - \bar{y}^2\right)} \qquad (7.5)$$

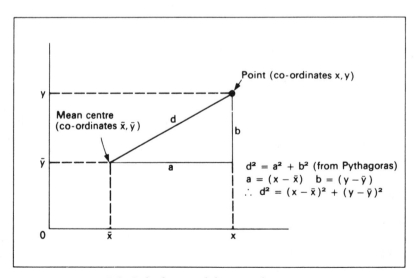

Figure 7.8 Calculation of distance from mean centre

Equation 7.5 is a direct algebraic transformation of Equation 7.4. The two equations should give exactly the same result if applied to a common set of data. Equation 7.5 is preferred simply because it reduces the number of steps involved in the calculation of standard distance.

Table 7.3 gives the calculation of the standard distance of the point pattern shown in Figure 7.7. In Figure 7.9 this standard distance has been used as the radius of a circle which provides a geographical representation of the dispersion of points about the mean centre.

One point to note about the calculation of standard distance is that it exaggerates the importance of the extreme points in a distribution, those furthest from the mean centre. As a result of squaring all the distances from the mean centre, the extreme points have a disproportionate influence on the value of the standard distance. The reader should therefore be aware, and beware, of the fact that standard distance is particularly sensitive to what might be considered 'rogue' points in the distribution. This is similar to the way in which extreme values influence the calculation of the standard deviation (Section 2.2).

Standard distance can also be calculated for weighted point data. This involved some changes to the equations given above. The modified equations are not given here, but the computer program for standard distance (listed at the end of the chapter) will handle weighted data.

Exercise 7.3

(a) Calculate the standard distance for the point pattern shown in Figure 7.3.
(b) Calculate the standard distance for the distribution of towns in Nottinghamshire (Figure 7.5).

Standard Deviational Ellipse

Standard distance is a convenient measure of dispersion in point patterns since it summarizes the spread of points in just one value. However,

Table 7.3 Calculation of Standard Distance

	Co-ordinates			
Point	x	y	x^2	y^2
A	1	2	1	4
B	2	3	4	9
C	2	2	4	4
D	2	1	4	1
E	3	4	9	16
F	3	3	9	9
G	3	1	9	1
H	4	4	16	16
n = 8	20	20	56	60

$$\bar{x} = \frac{20}{8} = 2.5, \bar{y} = \frac{20}{8} = 2.5$$

$$\text{standard distance} = \sqrt{\left(\frac{\sum x^2}{n} - \bar{x}^2\right) + \left(\frac{\sum y^2}{n} - \bar{y}^2\right)}$$

$$= \sqrt{\left(\frac{56}{8} - 2.5^2\right) + \left(\frac{60}{8} - 2.5^2\right)} = 1.414$$

it takes no account of the fact that the spread about the mean centre may be different in different **directions**. In Figure 7.9, for example, standard distance describes dispersion in terms of a circle about the mean centre. From the figure, however, it is apparent that dispersion is

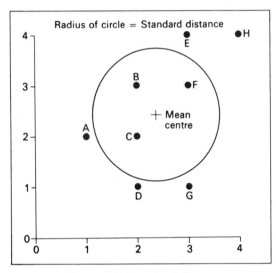

Figure 7.9 Standard distance

greater in a northeast-southwest direction than it is in a southeast-northwest direction.

The *standard deviational ellipse* as the name implies, is a measure which summarizes dispersion in a point pattern in terms of an ellipse rather than a circle. The ellipse is centred on the mean centre, with its long axis in the direction of maximum dispersion and its short axis in the direction of minimum dispersion. It can be shown mathematically that the axis of maximum dispersion in a point pattern is always at right angles to the axis of minimum dispersion, so that an ellipse is an appropriate measure.

In order to fit an ellipse about the mean centre of a point pattern it is necessary to know:

(1) the length of the short axis,
(2) the length of the long axis,
(3) the orientation of the ellipse.

The steps involved in fitting a standard deviational ellipse to a point pattern are shown in Figure 7.10. The mathematical calculations required for these steps can be summarized in a

few equations. To *transpose* the co-ordinate system:

$$x' = x - \bar{x}, \qquad y' = y - \bar{y} \qquad (7.6)$$

i.e. subtract the mean from each of the original x and y co-ordinates to give the transposed x and y co-ordinates, denoted by x' ('x prime') and y' ('y prime').

To calculate the angle of rotation:

$$\tan \theta = \frac{(\sum x'^2 - \sum y'^2) + \sqrt{(\sum x'^2 - \sum y'^2)^2 + 4(\sum x'y')^2}}{2 \sum x'y'} \qquad (7.7)$$

where $\tan \theta$ is the tangent of the angle of rotation, and x' y' are the transposed x and y co-ordinates.
To calculate the standard deviation along the x axis of the ellipse:

$$\sigma_x = \sqrt{\frac{\sum (x' \cos \theta - y' \sin \theta)^2}{n}} \qquad (7.8)$$

where σ_x is the standard deviation parallel to the x axis of the ellipse, θ is the angle of rotation, x' and y' are the transposed co-ordinates of the points, and n is the number of points. Equation 7.8 can be modified to give an equation which, although algebraically identical, involves fewer steps in calculation:

$$\sigma_x = \sqrt{\frac{(\sum x'^2) \cos^2 \theta - 2(\sum x'y') \sin \theta \cos \theta + (\sum y'^2) \sin^2 \theta}{n}} \qquad (7.9)$$

where $\cos^2 \theta$ is the conventional way of writing $(\cos \theta)^2$, in other words the square of the cosine of the angle. Similarly $\sin^2 \theta$ is the same as $(\sin \theta)^2$. Equation 7.9, although apparently more complex, is considerably faster to use than Equation 7.8, since it avoids the necessity of multiplying each x' and y' by $\cos \theta$ and $\sin \theta$ individually.
To calculate the standard deviation along the y axis of the ellipse:

$$\sigma_y = \sqrt{\frac{\sum (x' \sin \theta + y' \cos \theta)^2}{n}} \qquad (7.10)$$

or, for faster calculation:

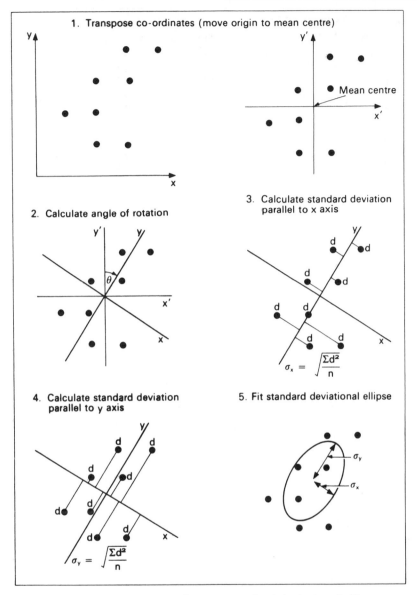

Figure 7.10 Steps in fitting a standard deviational ellipse

$$\sigma_y = \sqrt{\frac{(\sum x'^2)\sin^2\theta + 2(\sum x'y')\sin\theta\cos\theta + (\sum y'^2)\cos^2\theta}{n}}$$

(7.11)

The application of these equations to the point pattern shown in Figure 7.7 is given in Table 7.4, and the resultant standard deviational ellipse is fitted to the data in Figure 7.11. The reader should note that the lengths of the two axes are given by $2\sigma_x$ and $2\sigma_y$. It is also important to note that the angle θ given by Equation 7.7 is the angle between the transposed y axis and the y axis of the ellipse. The angle is measured **clockwise** from the transposed y axis, like a bearing from north on a map (see, for example Figure 7.11). Equation 7.7 can produce a negative value of tan θ. In this case the negative sign is ignored when looking up the angle in

Table 7.4 *Fitting a Standard Deviational Ellipse: 1*

| Point | Original | | Transposed ($x' = x - \bar{x}$, $y' = y - \bar{y}$) | | | | |
	x	y	x'	y'	x'^2	y'^2	x'y'
A	1	2	−1.5	−0.5	2.25	0.25	0.75
B	2	3	−0.5	0.5	0.25	0.25	−0.25
C	2	2	−0.5	−0.5	0.25	0.25	0.25
D	2	1	−0.5	−1.5	0.25	2.25	0.75
E	3	4	0.5	1.5	0.25	2.25	0.75
F	3	3	0.5	0.5	0.25	0.25	0.25
G	3	1	0.5	−1.5	0.25	2.25	−0.75
H	4	4	1.5	1.5	2.25	2.25	2.25

$n = 8$ $\sum x = 20$ $\sum y = 20$ $\sum x' = 0.0$ $\sum y' = 0.0$ $\sum x'^2 = 6.00$ $\sum y'^2 = 10.00$ $\sum x'y' = 4.00$

$\bar{x} = 2.5$ $\bar{y} = 2.5$

$$\tan \theta = \frac{(\sum x'^2 - \sum y'^2) + \sqrt{(\sum x'^2 - \sum y'^2)^2 + 4(\sum x'y')^2}}{2 \sum x'y'}$$

$$= \frac{(6.0 - 10.0) + \sqrt{(6.0 - 10.0)^2 + 4(4.0)^2}}{2 \times 4.0} = \frac{-4.0 + \sqrt{(-4.0)^2 + 4(16)}}{8}$$

$$= \frac{-4.0 + \sqrt{16.00 + 64}}{8} = \frac{-4.0 + \sqrt{80.00}}{8} = \frac{-4.0 + 8.94427}{8}$$

$$= \frac{4.94427}{8} = 0.61803 \qquad \therefore \ \theta = 31° \ 43'$$

$\sin \theta = 0.5257$ $\cos \theta = 0.8507$ $\sin^2 \theta = 0.2764$ $\cos^2 \theta = 0.7237$

$\sin \theta \cos \theta = 0.4472$

$$\sigma_x = \sqrt{\frac{(\sum x'^2) \cos^2 \theta - 2(\sum x'y') \sin \theta \cos \theta + (\sum y'^2) \sin^2 \theta}{n}}$$

$$= \sqrt{\frac{(6.0 \times 0.7237) - (2 \times 4.0)0.4472 + (10.0 \times 0.2764)}{8}}$$

$$= \sqrt{\frac{4.3432 - 3.5776 + 2.764}{8}} = \sqrt{\frac{3.5286}{8}} = \sqrt{0.4411} = 0.664$$

$$\sigma_y = \sqrt{\frac{(\sum x'^2) \sin^2 \theta + 2(\sum x'y') \sin \theta \cos \theta + (\sum y'^2) \cos^2 \theta}{n}}$$

$$= \sqrt{\frac{(6.0 \times 0.2764) + (2 \times 4.0)0.4472 + (10.0 \times 0.7237)}{8}}$$

$$= \sqrt{\frac{1.6584 + 3.5776 + 7.237}{8}} = \sqrt{\frac{12.4730}{8}} = \sqrt{1.5591} = 1.249$$

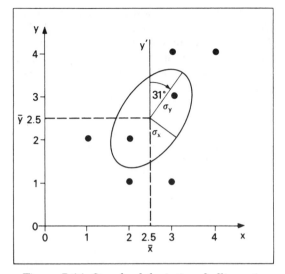

Figure 7.11 Standard deviational ellipse: 1

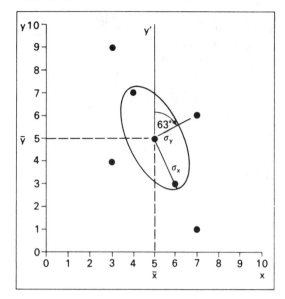

Figure 7.12 Standard deviational ellipse: 2

Table 7.5 Fitting a Standard Deviational Ellipse: 2

x	y	x'	y'	x'^2	y'^2	x'y'
3	4	−2	−1	4	1	2
3	9	−2	4	4	16	−8
4	7	−1	2	1	4	−2
5	5	0	0	0	0	0
6	3	1	−2	1	4	−2
7	1	2	−4	4	16	−8
7	6	2	1	4	1	2

$\sum x = 35 \quad \sum y = 35 \quad \sum x' = 0 \quad \sum y' = 0 \quad \sum x'^2 = 18 \quad \sum y'^2 = 42 \quad \sum x'y' = -16$
$\bar{x} = 5 \qquad \bar{y} = 5$

$$\tan \theta = \frac{(\sum x'^2 - \sum y'^2) + \sqrt{(\sum x'^2 - \sum y'^2)^2 + 4(\sum x'y')^2}}{2\sum x'y'}$$

$$= \frac{(18 - 42) + \sqrt{(18 - 42)^2 + 4(-16)^2}}{2(-16)}$$

$$= \frac{-24 + \sqrt{(-24)^2 + 4(256)}}{-32} = \frac{-24 + \sqrt{576 + 1024}}{-32} = \frac{-24 + \sqrt{1600}}{-32}$$

$$= \frac{-24 + 40}{-32} = \frac{16}{-32} = -0.5 \quad \therefore \theta = 90° - 26° \, 34' = 63° \, 26'$$

tables of tangents, but the angle found from tables must be subtracted from 90° in order to give the correct value of θ.

Table 7.5 gives the calculation of the angle of rotation for the point pattern shown in Figure 7.12. The tangent produced by the equation is negative (-0.5). The negative sign is ignored when using the tangent tables, so the angle found is 26° 34′. This angle is then subtracted from 90° to give the correct value of θ, which is 63° 26′. The two standard deviations can now be calculated in the usual way:

$$\sin \theta = 0.89451 \quad \cos \theta = 0.44828 \quad \sin^2\theta = 0.80015$$
$$\cos^2\theta = 0.20095 \quad \sin \theta \cos \theta = 0.40077$$

$$\sigma_x = \sqrt{\dfrac{(\sum x'^2)\cos^2\theta - 2(\sum x'y')\sin \theta \cos \theta + (\sum y'^2)\sin^2 \theta}{n}}$$

$$= \sqrt{\dfrac{(18)0.20095 - 2(-16)0.40077 + (42)0.80015}{7}}$$

$$= \sqrt{\dfrac{50.048}{7}} = \sqrt{7.150} = 2.674$$

$$\sigma_y = \sqrt{\dfrac{(\sum x'^2)\sin^2 \theta + 2(\sum x'y')\sin \theta \cos \theta + (\sum y'^2)\cos^2\theta}{n}}$$

$$= \sqrt{\dfrac{(18)0.80015 + 2(-16)0.40077 + (42)0.20095}{7}}$$

$$= \sqrt{\dfrac{10.018}{7}} = \sqrt{1.431} = 1.196$$

The fitted standard deviational ellipse is shown in Figure 7.12.

Exercise 7.4

Calculate θ, σ_x and σ_y for the point pattern shown in Figure 7.1. Remember that the coordinates must first be translated ($x \to x'$, $y \to y'$) before the angle of rotation can be calculated.

So far various descriptive measures for spatial distributions have been discussed and illustrated with rather simple examples. To give an idea of the usefulness of these measures in actual geographical research, the following paragraphs describe their application to an investigation of students' 'mental maps' of the city of Nottingham.

A class of first year geography students at Nottingham University was given a base map showing two well known landmarks by which the students could orientate themselves. They were then asked to mark the positions of 20 other places, without reference to any other maps or plans of Nottingham. A number of hypotheses was to be tested using the data collected. The sample of students taking part in the exercise was broken down into sub-samples according to various criteria. These were home background (whether predominantly urban or predominantly rural), whether they used private or public transport when travelling around the city, and whether they lived on or off the campus. It was hypothesized that there would be differences between the sub-samples in terms of each group's 'average perception' of the location of the various places. For example, it might be thought that people with a predominantly urban background would have less difficulty adjusting to a new city environment than those from a rural area and would therefore have more accurate spatial perceptions. Another hypothesis was that places further from the University would be less accurately located by **all** members of the sample than nearer places.

In testing these hypotheses the application of descriptive spatial statistics was an essential preliminary. The mean centre of all the perceived locations for any place provided an 'average' perceived location for the sample, or for any sub-sample. Distance and direction of this mean centre from the true location of the place provided a measure of accuracy of perception and enabled any systematic 'mis-location' of places to be quantified. Several places were consistently perceived as being considerably nearer to the University than they actually are. Standard distance relative to the mean centre provided a concise measure of the extent of agreement between members of the sample or sub-sample about the location of a particular place. In many cases there was a considerable amount of agreement about the location of a place (small standard distance), even though all the students involved had misplaced it by a kilometre or more from its true location!

Use of the standard deviational ellipse enabled any consistent spatial bias in perceived location to be measured. The perceived locations of a place which was known, or believed, to be due west of the University might be expected to lie within a standard deviational ellipse with its long axis orientated east-west. In other words, there would be more error in an east-west direction than in a north-south direction. Figure 7.13 shows some of the ways in which descriptive spatial statistics were used in analysing the results of the perception exercise.

The standard deviational ellipse can also be calculated for weighted point data. This involves some changes to the equations given above. The modified equations are not given here, but the computer program for the standard deviational ellipse (listed at the end of the chapter) will handle weighted data.

7.3 Shape

The measurement of shape is an obvious area in which the statistician can help the geographer. However, although shape is a very basic concept of which the human brain, and indeed that of other animals, has an intuitive appreciation, it is extremely difficult to quantify. Visual images processed by the brain tend to be sorted into a number of shape categories. Elementary shapes such as circles, triangles, squares and rectangles are recognized, and more complex shapes are appreciated in relation to these 'primitive' categories.

All of this is not very helpful to a geographer looking for a concise statistic which can be used to measure shape as a continuous variable. The fact is that only certain specific characteristics of shape can be quantified. It is possible to say how circular a shape is, or how square or how triangular, but not to say that a particular geographical area has a shape of so many 'shape units'.

The most commonly measured characteristic of shape is *compactness*. This is effectively a

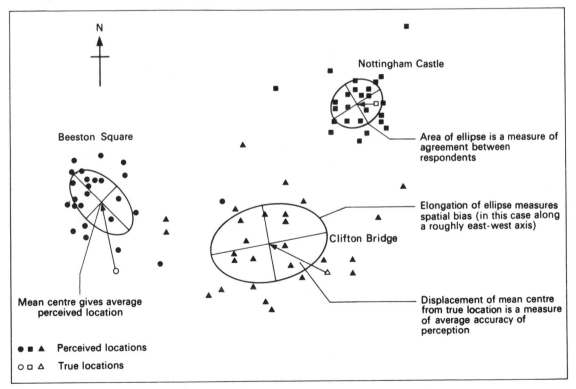

Figure 7.13 Descriptive spatial statistics applied to a perception study

measure of how far a shape deviates from the most compact possible shape, a circle. A circle is the most compact shape in the sense that it has the smallest possible perimeter relative to the area contained within it. Thus one measure of compactness would be the ratio of the length of the perimeter of a shape to its area. This is an index suggested by Pounds (1963).

$$S_1 = \frac{P}{A} \qquad (7.12)$$

where S_1 is the first shape index (this notation is merely used for convenience in this book, it is not 'official'), P is the length of the perimeter, and A is the area.

Although it is possible to calculate the length of the perimeter and the area for any perfect geometrical shape, it is quite difficult to do this accurately for irregular geographical units. Area is often known from official sources, particularly in the case of administrative units, but it is very difficult to measure the length of the boundary (perimeter) of an irregular unit. Working from maps at different scales it is possible to obtain widely differing estimates of the length of the boundary. In any case the use of the perimeter in a shape index makes that index oversensitive to the sinuosity of the perimeter. An area with a highly indented perimeter, such as an English county, will have a far higher value of S_1 than an area with a smooth outline, such as one of the Midwestern states of the U.S.A., although to the eye their shapes may be very similar.

A more serious problem is that the value of the index is not independent of either units of measurement or the absolute size of the shape concerned. A large circle will have a different value of S_1 from that of a small circle, even if their areas and perimeters are measured in the same units. Conversely, two identical circles will yield different values of S_1 if one is measured in feet and square feet and the other in metres and square metres. The reader can verify this by experiment.

Because of these two difficulties, S_1 is not a widely used shape index. It is discussed here to illustrate some of the problems in devising workable indices of shape. Four other fairly straightforward shape indices are illustrated in Figure 7.14. In Figure 7.15 the four indices are

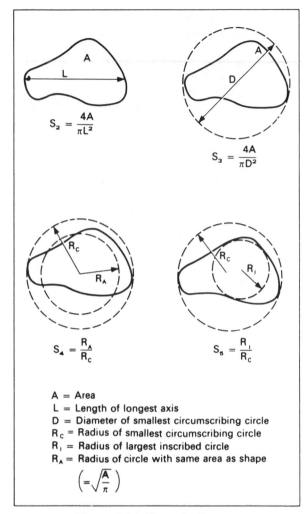

Figure 7.14 Shape indices

applied to some example shapes to show the range of values produced. Each of the indices is scaled in such a way that the most compact shape, a circle, has an index of 1.0. As the shapes become less compact the index falls in each case.

Exercise 7.5

Figure 7.1 shows various counties of Great Britain before and after boundary revision resulting from the reform of local government in 1974.

(a) Calculate S_2, S_3, S_4 and S_5 for each county before and after boundary revision.

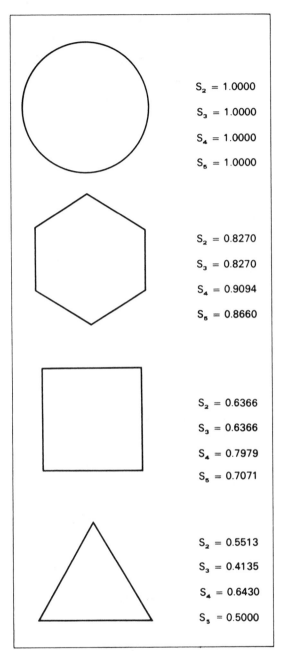

$S_2 = 1.0000$

$S_3 = 1.0000$

$S_4 = 1.0000$

$S_5 = 1.0000$

$S_2 = 0.8270$

$S_3 = 0.8270$

$S_4 = 0.9094$

$S_5 = 0.8660$

$S_2 = 0.6366$

$S_3 = 0.6366$

$S_4 = 0.7979$

$S_5 = 0.7071$

$S_2 = 0.5513$

$S_3 = 0.4135$

$S_4 = 0.6430$

$S_5 = 0.5000$

Figure 7.15 Shape indices

(b) Have the counties been made more compact as a result of the changes? As a possible class exercise, produce a histogram to show the range of values of the shape indices calculated by the class. Which index appears to be least sensitive to measurement errors?

Various other, more complicated shape indices have been devised, for example by Boyce and Clark (1964), Bunge (1966) and Taylor (1971). Each of these indices involves an arbitrary number of measurements and the value of the index for any given shape can vary considerably according to the number of measurements taken. For most practical purposes one of the indices described here should suffice, and at least permit rudimentary comparisons to be made between the shapes of different geographical units. It is perhaps worth repeating that shape is a very difficult property to define. It is even more difficult to measure. The shape indices described here are really measures of compactness, which is only one characteristic of shape.

7.4 Pattern

It can be argued that the existence of *pattern* in the spatial arrangement of phenomena on the earth's surface provides a fundamental stimulus to much of the geographer's work. Geographers talk of 'settlement patterns', 'land use patterns', 'drainage patterns'. In each case 'pattern' implies some sort of spatial regularity, which in turn is taken as a sign of the workings of a regular process.

The recognition and measurement of pattern is therefore of great importance to the geographer, although techniques for this are as yet poorly developed. In this section one widely used technique for the measurement of pattern will be discussed. This example has been chosen partly because it is fairly simple and already in common use in geography, but also because it is widely misused.

Nearest-neighbour analysis

Nearest-neighbour analysis is a technique, developed by plant ecologists (Clark and Evans, 1954), which is specifically designed for measuring pattern in terms of the arrangement of a set of points in two, or indeed three, dimensions. In essence the technique is quite straightforward. It involves calculating the

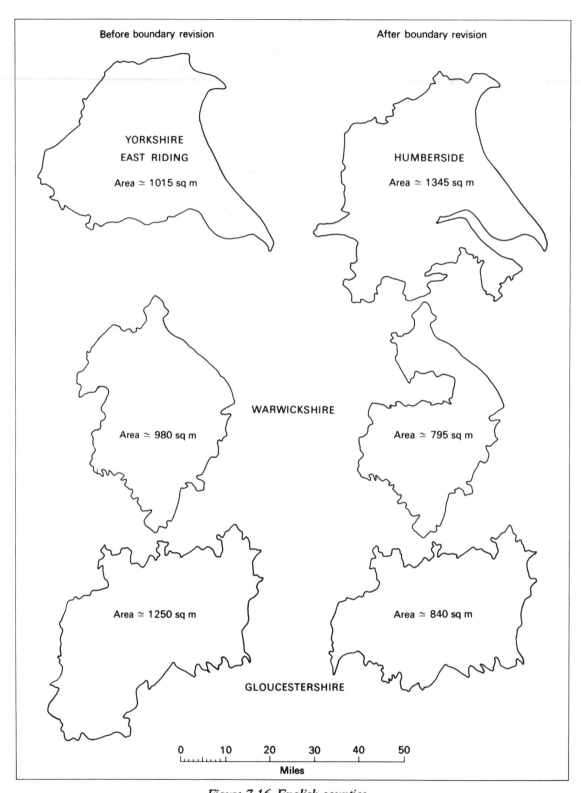

Figure 7.16 English counties

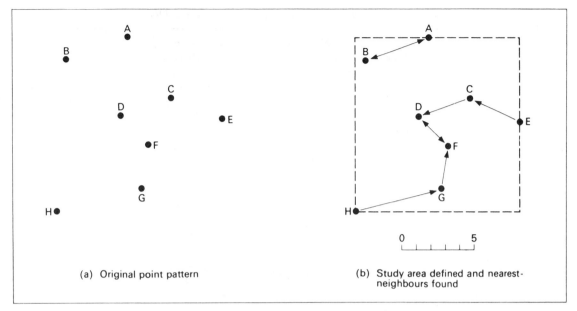

(a) Original point pattern

(b) Study area defined and nearest-neighbours found

Figure 7.17 Nearest-neighbour analysis

mean of the distances between all points and their nearest neighbours.

Figure 7.17a shows the distribution of a set of points within an area. At this stage it does not matter what these points represent, although it will become apparent that the interpretation of nearest-neighbour analysis depends very much on the phenomenon under consideration. In Figure 7.17b each point has been linked by an arrow to its nearest neighbour. An arbitrary study area has also been superimposed as a rectangle which encloses all the points as closely as possible. This is a convention which has been adopted by many users of nearest-neighbour analysis. In some cases, such as points A and B, nearest neighbours form a reflexive pair, i.e. each one is the nearest neighbour of the other. Once the nearest neighbours have been identified, the nearest-neighbour distance for each point can be measured, and the *mean nearest-neighbour distance* found. The number of nearest-neighbour distances is always the same as the number of points. These calculations are given for Figure 7.17 in Table 7.6. It can be seen that the mean nearest-neighbour distance in this case is 4.125 km.

It is possible to calculate theoretical values of the mean nearest-neighbour distance for cer-

tain types of spatial arrangements of points. The theoretical, or *expected*, mean nearest-neighbour distance for a random arrangement can be calculated from the following equation:

$$\bar{d}_{ran} = \frac{1}{2\sqrt{p}} \qquad (7.13)$$

where $\bar{d}_{ran}$ is the expected mean nearest-neighbour distance for a random arrangement of points, p is the density of points per unit area (the number of points divided by the area). The more mathematically inclined reader should consult Clark and Evans (1954) for the derivation of this equation.

In this example the area is 144 km^2 (12 km by 12 km), and there are 8 points. The density of points is therefore $8/144 = 0.056$. Note that the area **must** be calculated in the same units, in this case kilometres, as the nearest-neighbour distances. Using Equation 7.13 the expected mean nearest-neighbour distance for a random arrangement of 8 points within an area of 144 km^2 can now be calculated:

$$\bar{d}_{ran} = \frac{1}{2\sqrt{0.056}} = \frac{1}{2 \times 0.237} = \frac{1}{0.474} = 2.11$$

If the points in Figure 7.17 had been arranged

Table 7.6 Calculation of Mean Nearest-Neighbour Distance

Point	Nearest neighbour	Nearest-neighbour distance d
A	B	5
B	A	5
C	D	4
D	F	3
E	C	4
F	D	3
G	F	3
H	G	6

n = 8	$\sum d = 33$

Mean nearest-neighbour distance: $\bar{d} = \dfrac{\sum d}{n} = \dfrac{33}{8} = 4.125$

randomly the mean nearest-neighbour distance could be expected to be 2.11 km.

It is also possible to calculate the expected mean nearest-neighbour distance for these points if they had been arranged in the most dispersed way possible. This arrangement is shown in Figure 7.18. The points have the maximum possible distance separating them; they

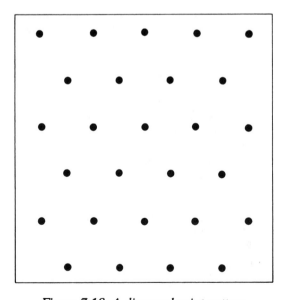

Figure 7.18 A dispersed point pattern

are as far apart as they can possibly be. This type of pattern is often referred to in descriptions of nearest-neighbour analysis as 'regular' or 'uniform'. It is felt that 'dispersed' more accurately describes the primary characteristic of this sort of pattern. There are many other types of pattern which could logically be described as 'regular' or 'uniform' but which do not exhibit the same degree of dispersal.

For a dispersed pattern it can be shown that the expected nearest-neighbour distance is given by:

$$\bar{d}_{dis} = \frac{2^{\frac{1}{2}}}{3^{\frac{1}{4}}\sqrt{p}} \quad \text{or} \quad \frac{1.07453}{\sqrt{p}} \quad (7.14)$$

For this example $\sqrt{p}$ has already been calculated as 0.237. The expected mean nearest-neighbour distance for a dispersed pattern in this case is therefore $1.07453/0.237 = 4.534$.

A third type of pattern can be considered in which the points are as close as possible to their nearest-neighbours. In the most extreme case of such a 'clustered' pattern each point will be so close to its nearest neighbour that the mean nearest-neighbour distance will be zero. Nearest-neighbour analysis therefore is concerned with finding the position of an observed spatial arrangement of points along a scale of pattern types. At the two extremes there are clustered and dispersed patterns, with a random arrangement falling somewhere in between.

In the present example, with 8 points in an area of 144 km^2, the expected mean nearest-neighbour distance is 0.0 km for a clustered pattern and 4.534 km for a dispersed pattern. The expected mean nearest-neighbour distance for a random arrangement is 2.11 km. Since the observed mean nearest-neighbour distance is 4.125 km. it can be said that the observed arrangement is close to being a dispersed pattern. The points are almost as far apart as they possibly could be.

The *nearest-neighbour index* provides a more concise measure of pattern in terms of a single value. The nearest-neighbour index is simply the observed mean nearest-neighbour distance divided by the expected mean nearest-neighbour distance for a random arrangement:

$$R = \frac{\bar{d}_{obs}}{\bar{d}_{ran}} \qquad (7.15)$$

where R is the nearest-neighbour index, $\bar{d}_{obs}$ is the observed mean nearest-neighbour distance, and $\bar{d}_{ran}$ is the expected mean nearest-neighbour distance for a random arrangement of points. The nearest-neighbour index can have a value between 0.0, indicating a completely clustered pattern, and 2.15, indicating a completely dispersed pattern. A random arrangement is indicated by a nearest-neighbour index of 1.0. In the present example:

$$R = \frac{4.125}{2.11} = 1.955$$

This suggests that the observed arrangement of points is very dispersed.

It is possible to apply a significance test to nearest-neighbour analysis to decide how probable it is that the observed arrangement of points occurred by chance. The null hypothesis is that the observed arrangement is the result of points being located at random within the study region. The test statistic used is based on the difference between $\bar{d}_{obs}$ and $\bar{d}_{ran}$, and is very similar in form to Student's t (Section 4.4):

$$c = \frac{\bar{d}_{obs} - \bar{d}_{ran}}{SE_{\bar{d}}} \qquad (7.16)$$

where c is the test statistic, $\bar{d}_{obs}$ is the observed mean nearest-neighbour distance, $\bar{d}_{ran}$ is the

expected mean nearest-neighbour distance for a random arrangement, and $SE_{\bar{d}}$ is the standard error of the mean nearest-neighbour distance.

The *standard error of the mean nearest-neighbour distance* is exactly analogous to the ordinary standard error of the mean (Section 3.2). It is calculated as follows:

$$SE_{\bar{d}} = \frac{0.26136}{\sqrt{np}} \qquad (7.17)$$

where n is the number of points and p is the density of points per unit area. For the arrangement shown in Figure 7.17 this gives:

$$SE_{\bar{d}} = \frac{0.26136}{\sqrt{8 \times 0.056}} = \frac{0.26136}{\sqrt{0.448}} = \frac{0.26136}{0.669} = 0.391$$

The test statistic c can now be calculated for Figure 7.17:

$$c = \frac{4.125 - 2.11}{0.669} = \frac{2.015}{0.669} = 3.012$$

The sampling distribution of the test statistic c is in fact the normal distribution. In other words, c is a *standard normal deviate*. Positive values denote a dispersed pattern and negative values a clustered one. Since c behaves like any other standard normal deviate (Section 1.3) there is a probability of 0.682 of obtaining a value of c between -1.0 and $+1.0$, and a probability of 0.954 of obtaining a value of c between -2.0 and $+2.0$ under the null hypothesis.

The significance of c can be checked by reference to tables of critical values of a standard normal deviate (Appendix C10). In the present case, assuming a significance level of 0.05, the appropriate critical value for a one-tailed test is 1.645, which is less than the observed value of c. The test therefore shows that the null hypothesis can be rejected at the 0.05 level, i.e. that the arrangement shown in Figure 7.17 can be considered to be 'significantly dispersed' at the 0.05 level.

It is possible to test the significance of results from nearest-neighbour analysis in another way, which avoids the necessity of calculating the value of c. For this purpose tables of critical values of the nearest-neighbour index R can be constructed (Appendix C11). Details of how these critical values are produced are given in Ebdon (1976). All that matters here is that the tables are used in much the same way as other tables of critical values. In the case of a dispersed pattern (R > 1.0) the calculated value of R needs to be as large as or larger than the critical value in order to be able to reject the null hypothesis at the chosen significance level. For a clustered pattern (R < 1.0) the calculated R must be less than or equal to the critical value. The only other value which needs to be known is the number of points in the pattern, n.

For the arrangement shown in Figure 7.17 the calculated value of R is 1.955, n is 8, and the chosen significance level is 0.05. The appropriate critical value of R for a dispersed pattern is found in Appendix C11a in the row labelled 8 and the column headed 0.05. The critical value, which can be denoted as $R_{dis}(0.05)$, is therefore 1.304. Since the calculated R is larger than the critical value, the null hypothesis of randomness can be rejected. The result of applying this form of significance test should, of course, always be the same as if c had been calculated.   In each of the examples discussed above a one-tailed test is appropriate since a direction of departure from randomness is specified, either towards clustering or towards dispersion. If the test is simply being used to determine whether or not a pattern can be considered to be random, a two-tailed test will be used. Critical values of R for a two-tailed test are given in Appendix C11b.

Exercise 7.6

Calculate the nearest-neighbour statistic for the arrangement of points shown in Figure 7.19. Is this pattern 'significantly clustered' at the 0.01 level?

Some Problems in the Application of Nearest-Neighbour Analysis

The various theoretical equations used in nearest-neighbour analysis, particularly 7.13 and 7.17, are based on two crucial assumptions. The first is that the points are located within an infinite area. The second is that the points are free to locate anywhere within that area. In most geographical situations these assumptions are manifestly unrealistic. In many cases where nearest-neighbour analysis is applied the

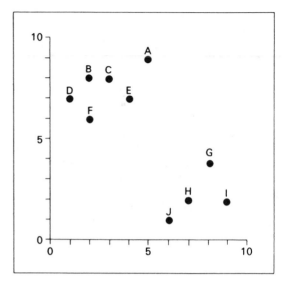

Figure 7.19 Point pattern for exercise 7.6

delimitation of the study area is quite arbitrary. Since the size of the area influences the point density, p, it also seriously affects the outcome of the analysis. If a study area is enlarged, without bringing more points into it, the effect is to increase the expected value of $\bar{d}_{ran}$, without of course changing the value of $\bar{d}_{obs}$. The nearest-neighbour index, R, will consequently be decreased in value, although the appearance of the observed arrangement of points and its degree of clustering or dispersion will not have changed at all.

Because of this problem of study area delimitation, one should be very wary indeed of comparisons made between nearest-neighbour analysis results from different areas. There is also some evidence (Ebdon, 1976) that the significance test described above is consistently biased towards finding significantly dispersed patterns. Dawson (1975) has also pointed out that it is possible for arrangements of points which are very dissimilar to have identical mean nearest-neighbour distances. The existence of these problems does not mean that nearest-neighbour analysis should not be used by geographers. The technique provides a very useful descriptive measure of point patterns, particularly for quantifying the increase or decrease in dispersion or clustering of a pattern through time, provided the definition of the study area remains the same. However, comparisons between different areas based on nearest-neighbour analysis should be treated with great caution, as should the significance test.

Exercise 7.7

Consider the situation shown in Figure 7.20.

The map shows the distribution of houses up for sale in a residential area.

(a) Use nearest-neighbour analysis to decide whether the pattern of houses for sale is significantly clustered at the 0.05 level.

(b) More importantly, consider whether the technique can validly be applied to this situation.

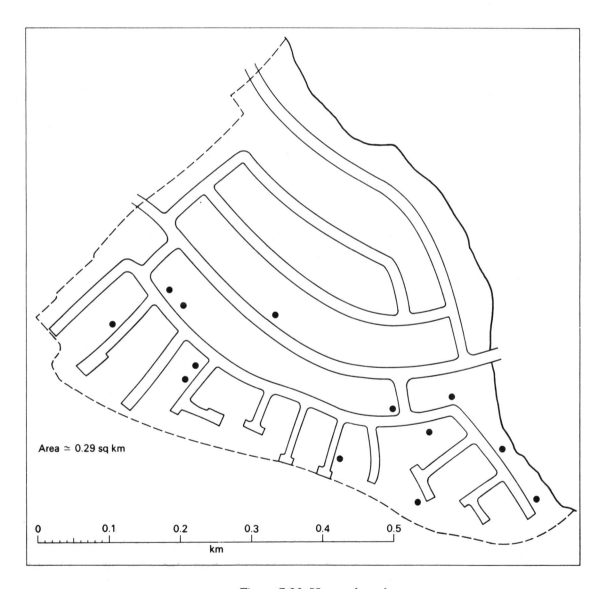

Area ≃ 0.29 sq km

Figure 7.20 Houses for sale

7.5 Spatial Relationships

The most recently developed, and possibly the most interesting, branch of geographical statistics is concerned with variation through space. The techniques which will be discussed in this section are used for measuring the way in which the values of a variable vary over a two-dimensional surface. Techniques are available for nominal, ordinal and interval data, and for data relating to points or areas. One technique of each type will be discussed here.

This branch of statistics is concerned with what is known as *spatial autocorrelation*. Autocorrelation (Section 6.1) is concerned with the relationship between successive values of residuals along a regression line. Strong autocorrelation means that successive values are strongly related; they vary in a systematic way. Spatial autocorrelation is a fairly simple extension of this concept into two dimensions. Strong spatial autocorrelation means that adjacent values or ones which are near to each other are strongly related. If values are simply arranged at random over the surface there should be no apparent spatial autocorrelation. In one sense the tests described in this section can be thought of as tests of whether the spatial arrangement of a set of values is or is not random. This is a logical extension of the concept of randomness in point patterns discussed in Section 7.4.

Join Count Statistics

The simplest form of spatial autocorrelation concerns the spatial arrangement of areas of two different types. Figure 7.21 shows three possible arrangements of 11 areas, 5 of which are of one type and the remaining 6 of another. For simplicity, and to comply with accepted practice in other statistical texts, these two types will be referred to as 'black' and 'white'.

In Figure 7.21a the five white areas are clustered together, in Figure 7.21b they are dispersed, and in Figure 7.21c no particular pattern is evident. These three spatial arrangements could be termed 'clustered', 'dispersed' and 'random' respectively. Note that in each case the underlying spatial structure is the same. It can be seen that there is a close analogy with

the three types of point pattern discussed in Section 7.4. A simple and quick method of quantifying the degree of clustering or dispersion, in fact to measure the spatial autocorrelation, in these spatial arrangements is to count the number of joins between black and white areas. If like areas are clustered there will be relatively few black/white joins; if they are dispersed there will be relatively many. A random arrangement of areas will produce an intermediate number of black/white joins.

At this stage it may be helpful to demonstrate the geographical applications of join counts. The counts can be applied to the spatial distribution of any variable which is measured on, or can be converted to, a **nominal** scale. The join count measures discussed here are only applicable to nominal variables consisting of two categories, i.e. *binary* variables. However, the technique can be extended to variables comprising more than two categories. This point will be discussed in more detail later. Some of the variables geographers deal with are inherently binary. In electoral geography, for example, interest might lie in the spatial distribution of 'socialist' constituencies compared with 'others'. The spatial arrangement of arable and non-arable farms, or parishes exhibiting population growth or decline could also be measured by join counts.

Other variables can be converted very readily into binary form. Medical geographers, for example, are interested in the spatial patterns of mortality from various diseases. Using join count measures it is possible to discover whether there is spatial autocorrelation amongst counties or hospital regions with above-average or below-average mortality rates for any particular disease. The essential requirements for the application of join counts are:

(1) data relating to **areas** (not points or lines),
(2) data measured in **binary** form.

SIGNIFICANCE

Counting the number of black/white joins is a simple task. However, in order to test the significance of spatial autocorrelation it is necessary to know how probable it is that the observed number of black/white joins could

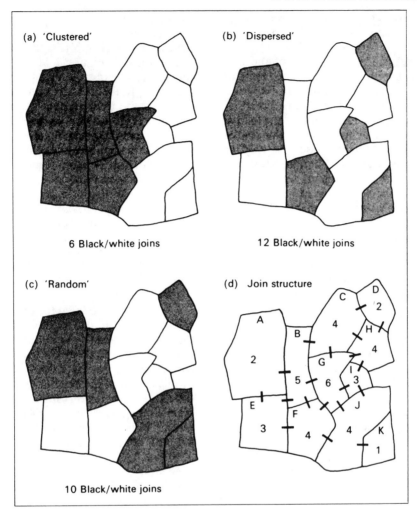

(a) 'Clustered'

6 Black/white joins

(b) 'Dispersed'

12 Black/white joins

(c) 'Random'

10 Black/white joins

(d) Join structure

Figure 7.21 Join count statistics

occur by chance. In order to estimate this probability it is necessary to know the expected number of black/white joins for a random pattern, and the standard deviation of that expected number. There is again an analogy here with the significance test used in nearest-neighbour analysis (Section 7.4).

A slight complication arises from the fact that there are two alternative forms of null hypothesis for the significance test:

(*a*) *Free Sampling* (or sampling with replacement)

This form of null hypothesis is only applicable if the probability that an area will be black or white is known without reference to the study

region (*a priori*). For example, in a study of 'socialist' compared with 'other' constituencies within a study region it is possible to calculate on the basis of national figures the probability of a constituency being 'socialist', or conversely 'non-socialist'.

(*b*) *Non-Free Sampling* (or sampling without replacement)

Under this form of null hypothesis only the situation within the study region is considered. No reference is made to any outside factors. This form of null hypothesis could also apply to the electoral geography example cited above. It might be considered that the national voting pattern could not be expected to apply within the particular study region. In this case it would be wrong to apply probabilities based on national figures.

In both cases the general form of the null hypothesis is that the spatial arrangement of black and white areas is random. The alternative hypothesis may be either directional or non-directional. A directional alternative hypothesis would specify whether the spatial arrangement was thought to be clustered or dispersed. A non-directional hypothesis would simply be that the arrangement was not random. As usual, the setting up of a directional alternative hypothesis means that a one-tailed test is appropriate. A non-directional alternative hypothesis implies a two-tailed test.

The choice of null hypothesis is an important consideration in the application of join counts. As a general guide it is probably wise to use the non-free sampling approach if there is any doubt whether probabilities calculated by reference to a larger area can be expected to apply within the study region. In all cases the assumptions made in adopting the non-free sampling null hypothesis are less extreme than those inherent in the free sampling approach. Once the form of null hypothesis has been chosen, the appropriate equations for the expected number of black/white joins and its standard deviation can be applied. The two approaches will be demonstrated in turn by applying them to the arrangements shown in Figure 7.21.

FREE SAMPLING

The equation for the expected number of black/white joins is:

$$E_{BW} = 2Jpq \qquad (7.18)$$

where E_{BW} is the expected number of black/white joins, J is the total number of joins in the study region, p is the probability that an area will be black, and q is the probability that an area will be white $(p + q = 1.0)$. The equation for the standard deviation is:

$$\sigma_{BW} = \sqrt{\begin{aligned}&[2J + \sum L(L-1)]pq \\ &-4[J + \sum L(L-1)]p^2q^2\end{aligned}} \qquad (7.19)$$

where σ_{BW} is the standard deviation of the number of black and white joins, J, p and q are as defined previously, and $\sum L(L-1)$ will be explained below.

The calculations involved in these two equations are quite straightforward and can best be demonstrated by applying the test to the situation shown in Figure 7.21a. The number of joins, J, can be obtained simply by counting. In Figure 7.21d the joins are marked, and it can be seen that the total number of joins is 19. This number of joins applies to all the different arrangements shown in Figure 7.21. A problem can occur in counting joins when two areas join at a point rather than along a line. In most situations it seems reasonable to consider this as a join, since the two areas are *contiguous* (touching) even though they do not actually share a border. It is up to the researcher to decide whether to include these joins. If one join of this type is included it is imperative that all such joins are counted, both in the total number of joins and in the number of black/white joins.

It will be assumed that the probabilities p and q in this case are 0.4 and 0.6. Remember that these would normally be calculated on the basis of the overall number of black and white areas within some area larger than the study region. In this case it might be that the total region, of which the study region forms a part, contains 40 black and 60 white areas, giving a total of 100 areas. The probabilities are therefore 40/100 or 0.4, and 60/100 or 0.6.

Figure 7.21d also shows the values of L. For each area L is the number of joins between that area and contiguous areas. The calculation of $\sum L(L-1)$ is shown in Table 7.7. As a check on the calculations so far, it will be seen that $\sum L$ should always be equal to 2J (twice the total number of joins).

All the necessary values have now been calculated for the situation in Figure 7.21a and can be substituted into the equations:

$$E_{BW} = 2 \times 19 \times 0.4 \times 0.6 = 9.12$$

$$\sigma_{BW} = \sqrt{(38 + 114)0.4 \times 0.6 - 4(19 + 114)0.4^2 \times 0.6}$$

$$= \sqrt{36.48 - 30.6432} = \sqrt{5.8368} = 2.416$$

If the observed number of black/white joins is defined as O_{BW}, the test statistic can now be calculated using the following equation:

$$z = \frac{O_{BW} - E_{BW}}{\sigma_{BW}} \qquad (7.20)$$

Table 7.7 Join Count Statistics: Calculation of $\sum L(L-1)$

Area	L	(L−1)	L(L−1)
A	2	1	2
B	5	4	20
C	4	3	12
D	2	1	2
E	3	2	6
F	4	3	12
G	6	5	30
H	4	3	12
I	3	2	6
J	4	3	12
K	1	0	0

$$\sum L = 38 \qquad \sum L(L-1) = 114$$

The test statistic z follows a normal distribution, i.e. it is a standard normal deviate. Once again there is a close parallel between z and the statistic c used in nearest-neighbour analysis (Section 7.4, Equation 7.16). To find the significance of a particular join count, the value of z can be compared with the appropriate critical value obtained from tables of a standard normal deviate (Appendix C10).

In the present example (Figure 7.21a):

$$z = \frac{6-9.12}{2.416} = -1.291$$

Adopting the 0.05 significance level, the critical value for a negative value of z is −1.960. A two-tailed test is appropriate in this case, since no expected direction of departure from randomness has been specified. In order to reject the null hypothesis (that the observed arrangement is random) the calculated z needs to be less than or equal to −1.960. The null hypothesis can not be rejected. It must be concluded that the observed arrangement could easily have occurred by chance under the null hypothesis. In other words the observed spatial arrangement of black and white areas in Figure 7.21a is not significantly different from a random arrangement.

This should seem to the reader to be contrary to common sense. The arrangement shown in Figure 7.21a has already been described as 'clustered'. However, in adopting the free sam-

pling null hypothesis the analysis takes account of circumstances outside the study region. Remember that the calculation of the probabilities p and q is based on a hypothetical overall situation in which there are 100 areas, of which 40 are black and 60 white. The significance test shows that, given this overall situation, it is not unlikely that a sample sub-region composed of 11 areas with a total of 19 joins could contain a cluster of 5 black areas by chance.

Exercise 7.8

(a) Calculate the value of z for the spatial arrangements shown in Figures 7.21b and c. Is either of these patterns significantly different from a random pattern at the 0.05 level? Assume that $p = 0.4$ and $q = 0.6$.

(b) Recalculate z for Figure 7.21a, this time assuming that $p = 0.45$ and $q = 0.55$. Is this arrangement 'significantly clustered' at the 0.05 level?

NON-FREE SAMPLING

Two equations need to be applied:

$$E_{BW} = \frac{2JBW}{n(n-1)} \qquad (7.21)$$

where E_{BW} is the expected number of black/white joins, J is the number of joins in the study region, B is the number of black areas and W the number of white areas in the study region, and n is the total number of areas in the study region ($n = B + W$).

$$\sigma_{BW} = \sqrt{ \begin{array}{c} E_{BW} + \dfrac{\sum L(L-1)BW}{n(n-1)} \\ + \dfrac{4[J(J-1)-\sum L(L-1)]B(B-1)W(W-1)}{n(n-1)(n-2)(n-3)} \\ - E_{BW}^2 \end{array} }$$

$$(7.22)$$

where $\sum L(L-1)$ is calculated in the same way as it is under the free sampling null hypothesis. For the situation shown in Figure 7.21a the following values apply: $J = 19$, $B = 5$, $W = 6$,

$n = 11$, and $\sum L(L-1) = 114$. These values can be substituted into the two equations:

$$E_{BW} = \frac{2 \times 19 \times 5 \times 6}{11 \times 10} = \frac{1140}{110} = 10.3636$$

$$\sigma_{BW} = \sqrt{\begin{array}{l} 10.3636 + \dfrac{114 \times 5 \times 6}{11 \times 10} \\ + \dfrac{4[(19 \times 18) - 114] \times 5 \times 4 \times 6 \times 5}{11 \times 10 \times 9 \times 8} \\ - 10.3636^2 \end{array}}$$

$$= \sqrt{10.3636 + 31.0909 + 69.0900 - 107.4042}$$

$$= \sqrt{3.1403} = 1.7721$$

The test statistic is calculated in the same way as under the free sampling null hypothesis, using Equation 7.20:

$$z = \frac{O_{BW} - E_{BW}}{\sigma_{BW}} = \frac{6 - 10.3636}{1.7721} = -2.462$$

Adopting the 0.05 significance level, the critical value of a negative standard normal deviate (from Appendix C10) is -1.645. The calculated value of z is less than the critical value, so the null hypothesis can be rejected. The observed arrangement of black and white areas (Figure 7.21a) is very unlikely to have occurred by chance; it can be said to be 'significantly clustered' at the 0.05 level.

Exercise 7.9

Using the non-free sampling null hypothesis, are the spatial arrangements shown in Figure 7.21b and c significantly different from random at the 0.01 level?

It may be worth re-emphasizing the difference in approach between the two forms of null hypothesis. Free sampling effectively involves the assumption that each area within a study region has the probability p of being black and a probability q of being white. These probabilities are known without reference to the study region. A simple simulation exercise could be constructed to show how the hypothesis works. Imagine a study region comprising say 10 areas, with a probability of 0.25 that any area will be black and a probability of 0.75 that any area will be white. It is possible to assign colours to the areas within the study region in a random way by picking coloured balls from an urn. An urn containing 25 black balls and 75 white balls gives the correct probabilities. To each area in turn a colour is assigned by picking a ball from the urn, noting down its colour, and returning it to the urn. When 10 balls have been taken out and returned the 10 areas will all have been coloured. The number of black/white joins can now be counted and the simulation repeated. From a large number, say 1000, runs of the simulation a reasonable approximation to the theoretical sampling distribution of the number of black/white joins can be derived.

Note that this sampling distribution would apply only to the particular study region, or at any rate only to one which has the same number of areas, the same 'join structure' and the same overall probabilities p and q. Fortunately, however, Equations 7.18 and 7.19 enable critical values from this sampling distribution to be calculated relatively easily for any spatial arrangement of areas and pairing of probabilities.

The non-free sampling approach is rather easier to appreciate. Imagine a study region composed of say 20 areas, of which 8 are black and the remaining 12 are white. There is a large number of ways (in fact 125,970) in which the 8 black and 12 white areas could be arranged within the given spatial structure of the study region. By going through every one of these combinations and noting down the number of black/white joins produced by each one, the sampling distribution could be discovered. A rather less time-consuming procedure would be to put 8 black and 12 white balls into an urn, and to carry out a simulation as before. With non-free sampling, however, the balls would not be replaced each time. After say 100 runs of the simulation a good estimate of the sampling distribution of the number of black/white joins would be produced.

Exercise 7.10

Figure 7.22 shows the outcome of the 1976 presidential election in the United States (excluding Alaska, Hawaii and the District of Columbia). Shaded states are those in which

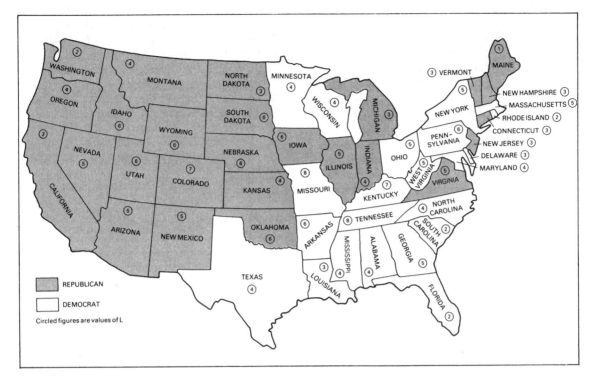

Figure 7.22 Popular Vote in the United States Presidential Election 1976

the Republican candidate gained a majority of the popular vote. The total number of votes cast in the U.S.A. as a whole was 40.831m for the Democratic candidate and 39.148m for the Republican candidate. The pattern of 'Republican States' certainly looks clustered, but is the clustering significant at the 0.01 level?

Spatial Autocorrelation Measures for Ordinal and Interval Data

In many geographical situations a lot of information is clearly lost if data are reduced to binary form. Several tests exist for measuring the spatial autocorrelation between ordinal or interval values relating to areas or points. One such measure has been devised by Moran (1950) and can be applied 'manually' to areas. It can also be applied to point data, but this application will be considered separately later.

For areal data the situation is very similar to that described for join counts except that the black and white categories are replaced by numerical values, either ranks or interval measurements. The equation for Moran's coefficient is:

$$I = \frac{n \sum_{(c)} (x_i - \bar{x})(x_j - \bar{x})}{J \sum (x - \bar{x})^2} \qquad (7.23)$$

where I is Moran's spatial autocorrelation coefficient, n is the number of areas in the study region, J is the number of joins, x is a value (ordinal or interval) for an area, $\bar{x}$ is the mean of all the values of the variable x, and x_i and x_j are the values for two contiguous areas (areas on either side of a join). The summation sign $\sum_{(c)}$ means that the value of $(x_i - \bar{x})(x_j - \bar{x})$ must be calculated for each pair of contiguous areas and all these values summed. The application of

Moran's coefficient can be demonstrated by a simple example.

Figure 7.23 shows a hypothetical study region composed of 6 areas, with 9 joins ($n = 6$, $J = 9$).

For each area the value of a particular variable, perhaps population density, has been measured, and these values are shown on the figure. The joins have been numbered to help clarify the calculations, which are set out in Table 7.8. The value of Moran's coefficient in this case is -0.183, although this value on its own is not of very much use in describing the degree of spatial autocorrelation in a variable. The range of possible values of I depends on the spatial structure of the particular study region. To find out what the value of I implies it is therefore necessary to carry out a significance test.

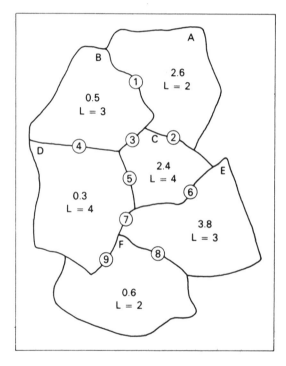

Figure 7.23 Spatial autocorrelation for areas

SIGNIFICANCE

As with the join count measures, the significance test involves calculating a standard normal deviate from the calculated value of I, the expected value of I and its standard deviation.

Once again there are two possible forms of null hypothesis: normality and randomization. These two forms are equivalent to free sampling and non-free sampling for binary data.

(a) *Normality* The null hypothesis is that the observed values of the variable x are the result of taking n values at random from a normally distributed population of values.

(b) *Randomization* The n values of the variable are taken as given, but the particular spatial arrangement of these values is considered in relation to all the ways in which the values could possibly be arranged within the study region. The question asked is: 'given the particular set of values of x, what is the probability that they could have been arranged in the observed way by chance?'

NORMALITY

The equation for the expected value of I under the null hypothesis of normality is:

$$E_I = -\frac{1}{n-1} \qquad (7.24)$$

The equation for the standard deviation of this value is:

$$\sigma_I = \sqrt{\frac{n^2 J + 3J^2 - n \sum L^2}{J^2(n^2-1)}} \qquad (7.25)$$

where σ_I is the standard deviation of I, n is the number of areas in the study region, and J is the number of joins. L is the number of areas to which an area is joined and $\sum L^2$ is the sum of the squares of all these values.

For the situation shown in Figure 7.23 the calculation of $\sum L^2$ is given in Table 7.8. The appropriate values can now be substituted into Equations 7.24 and 7.25:

$$E_I = -\frac{1}{6-1} = -0.2$$

$$\sigma_I = \sqrt{\frac{(6^2 \times 9) + (3 \times 9^2) - (6 \times 58)}{9^2(6^2-1)}}$$

$$= \sqrt{\frac{324 + 243 - 348}{81 \times 35}}$$

$$= \sqrt{\frac{219}{2835}} = \sqrt{0.077} = 0.278$$

Table 7.8 Calculation of Moran's Spatial Autocorrelation Coefficient I for Figure 7.23

Area	Joins L	Joins L^2	Area values x	Area values $(x - \bar{x})$	Area values $(x - \bar{x})^2$
A	2	4	2.6	0.9	0.81
B	3	9	0.5	−1.2	1.44
C	4	16	2.4	0.7	0.49
D	4	16	0.3	−1.4	1.96
E	3	9	3.8	2.1	4.41
F	2	4	0.6	−1.1	1.21

$n = 6$ $\sum L^2 = 58$ $\sum x = 10.2$ $\sum (x - \bar{x})^2 = 10.32$
$\bar{x} = 1.7$

Join Number	x_i	$(x_i - \bar{x})$	x_j	$(x_j - \bar{x})$	$(x_i - \bar{x})(x_j - \bar{x})$
1	0.5	−1.2	2.6	0.9	−1.08
2	2.6	0.9	2.4	0.7	0.63
3	0.5	−1.2	2.4	0.7	−0.84
4	0.5	−1.2	0.3	−1.4	1.68
5	0.3	−1.4	2.4	0.7	−0.98
6	2.4	0.7	3.8	2.1	1.47
7	0.3	−1.4	3.8	2.1	−2.94
8	3.8	2.1	0.6	−1.1	−2.31
9	0.3	−1.4	0.6	−1.1	1.54

$J = 9$ $\sum_{(c)} (x_i - x)(x_j - x) = -2.83$

$$I = \frac{n \sum_{(c)} (x_i - \bar{x})(x_j - \bar{x})}{J \sum (x - \bar{x})^2} = \frac{6 \times (-2.83)}{9 \times 10.32} = -0.183$$

The previously calculated value of I can now be converted into a standard normal deviate using the following equation:

$$z = \frac{I - E_I}{\sigma_I} \qquad (7.26)$$

In this case:

$$z = \frac{-0.183 - (-0.2)}{0.278} = \frac{0.017}{0.278} = 0.061$$

Note that the expected value of I for a random arrangement is small and negative, in this case −0.2. A smaller value, one further from zero in the negative direction, implies disper-

sion. Positive values of I imply clustering. This can be summarized in diagrammatic form:

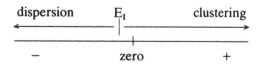

After converting the observed I to a standard normal deviate, its significance can be assessed by reference to a table of critical values (Appendix C10).

Adopting the 0.05 significance level, the two-tailed critical value for a positive standard normal deviate is 1.960. A two-tailed test is appropriate here, since no specific direction of departure from randomness has been suggested. The observed z (0.061) is less than the critical value, so the null hypothesis must be accepted. The observed arrangement of values in Figure 7.23 is not significantly different from random. It could easily have occurred under the null hypothesis of random sampling from a normally distributed population.

RANDOMIZATION

The equation for the expected value of I is the same as under the assumption of normality. In the case of Figure 7.23, then:

$$E_I = -\frac{1}{n-1} = -\frac{1}{5} = -0.2$$

The equation for the standard deviation, however, is rather more complex if the randomization null hypothesis is adopted:

$$\sigma_I = \sqrt{\frac{n[J(n^2+3-3n)+3J^2-n\sum L^2]}{J^2(n-1)(n-2)(n-3)}} \quad (7.27)$$

where, in addition to symbols already defined in this section, k is the kurtosis of the variable x. Kurtosis has been discussed in a previous chap-

ter (Section 2.3). It can be thought of as a measure of the 'peakiness' of the distribution of the values of x. To avoid repetition, the calculation of kurtosis for the values shown in Figure 7.23 is given in Table 7.9. All the values can now be substituted into Equation 7.27:

$$\sigma_I = \sqrt{\frac{6[9(6^2+3-18)+(3\times9^2)-(6\times58)]}{-1.56185[9(6^2-6)+(6\times9^2)-(2\times6\times58)]}{9^2(6-1)(6-2)(6-3)}}$$

$$= \sqrt{\frac{6(189+243-348)}{-1.56185(270+486-696)}{81\times5\times4\times3}}$$

$$= \sqrt{\frac{504-93.711}{4860}}$$

$$= \sqrt{0.08442} = 0.29055$$

Finally the observed value of I can be converted into a standard normal deviate in the usual way:

$$z = \frac{I-E_I}{\sigma_I} = \frac{-0.183-(-0.2)}{0.291} = \frac{0.017}{0.291} = 0.058$$

If the 0.01 significance level is adopted in this case, the two-tailed critical value of z is 2.576 (from Appendix C10). The calculated value of z is less than the critical value, so the null

Table 7.9 Calculation of Kurtosis for Randomization Significance Test of I

Area	x	$(x-\bar{x})$	$(x-\bar{x})^2$	$(x-\bar{x})^4$
A	2.6	0.9	0.81	0.6561
B	0.5	−1.2	1.44	2.0736
C	2.4	0.7	0.49	0.2401
D	0.3	−1.4	1.96	3.8416
E	3.8	2.1	4.41	19.4481
F	0.6	−1.1	1.21	1.4641

$\sum x = 10.2$ $\sum(x-\bar{x})^2 = 10.32$ $\sum(x-\bar{x})^4 = 27.7236$

$\bar{x} = 1.7$

$$\sigma = \sqrt{\frac{\sum(x-\bar{x})^2}{n}} = \sqrt{\frac{10.32}{6}} = \sqrt{1.72} = 1.31149$$

$$\text{kurtosis} = \frac{\sum(x-\bar{x})^4}{n\sigma^4} = \frac{27.7236}{6\times1.31149^4} = \frac{27.7236}{6\times2.9584} = \frac{27.7236}{17.7504} = 1.56185$$

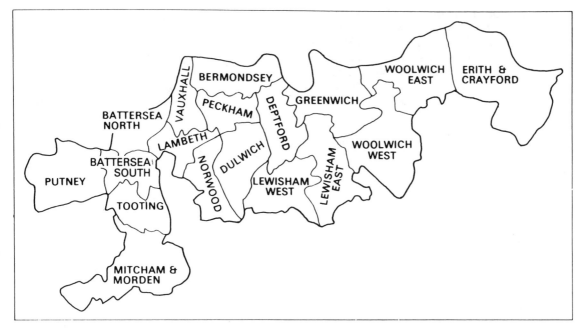

Figure 7.24 South London Labour constituencies: General Election Autumn 1974

hypothesis must be accepted. The observed arrangement is not significantly different from random. It could easily have occurred by chance under the null hypothesis of randomization.

Exercise 7.11

Figure 7.24 shows a group of constituencies (electoral districts) in South London. In each of them a Labour Member of Parliament was elected in the British General Election of Autumn 1974. Table 7.10 gives the percentage of the total vote polled by the Labour Party candidate in each constituency. Is there a significant degree of spatial autocorrelation between these percentages at the 0.05 level? Give some thought to the choice of null hypothesis to be used (normality or randomization) before starting the calculations. Try to write down a justification for your choice.

Further Topics in Spatial Autocorrelation

Earlier in this chapter it was suggested that join count measures could be applied to nominal data consisting of more than just two categories.

Table 7.10 Labour's Percentage Share of the Vote in the Autumn 1974 General Election (Selected South London Constituencies)

Battersea North	38.3
Battersea South	30.7
Bermondsey	41.4
Deptford	34.6
Dulwich	32.3
Erith & Crayford	37.4
Greenwich	36.3
Lambeth	31.6
Lewisham East	35.0
Lewisham West	33.8
Mitcham & Morden	34.2
Norwood	31.1
Peckham	39.1
Putney	32.5
Tooting	34.5
Vauxhall	33.3
Woolwich East	38.9
Woolwich West	34.8

One approach to this problem is to study the frequency distribution of areas with different numbers of joins of various types, and then apply a form of chi square test. The application of this form of spatial autocorrelation test is discussed in Cliff, Martin and Ord (1975) and Dacey (1968). Two other sources of information on spatial autocorrelation applied to nominal data comprising more than two categories are Cliff and Ord (1973), and Krishna Iyer (1949).

It was also stated earlier that spatial autocorrelation can be extended to situations involving point values. Here, instead of considering the relationship between pairs of contiguous area values, it is necessary to measure the relationship between **all** pairs of point values, taking into account the distance separating them. If there are n points there will be $n(n-1)/2$ possible pairs of points. With as few as 20 points this means that 190 pairs of values need to be multiplied and summed. With 50 points there will be 1,225 pairs! The technique can quite easily be computerised, but it is very tedious to try and apply it by hand to anything other than a very simple situation.

For the sake of completeness, however, and in the belief that geographers will soon need to understand this form of spatial autocorrelation, a very simple example will be given. A revised version of Moran's I is used:

$$I = \frac{n\sum_{(p)} w_{ij}(x_i - \bar{x})(x_j - \bar{x})}{(\sum_{(p)} w_{ij})\sum (x - \bar{x})^2} \quad (7.28)$$

where, in addition to previously defined notation, w_{ij} is the weight given to the relationship between two points i and j. The weight used is most commonly the reciprocal of the distance between the two points. If the distance between points i and j is defined as d_{ij}, then $w_{ij} = 1/d_{ij}$. The summation sign $\sum_{(p)}$ means that the value of $w_{ij}(x_i - \bar{x})(x_j - \bar{x})$ must be calculated for all pairs of values ($n(n-1)/2$ pairs) and summed. The calculation of I for the situation shown in Figure 7.3 is given in Table 7.11.

SIGNIFICANCE TEST

As with the I test for areal data, two forms of null hypothesis apply:

(a) *Normality*

$$E_I = -\frac{1}{n-1}$$

$$\sigma_I = \sqrt{\frac{n^2 A + 3B^2 - nC}{(n^2 - 1)B^2}} \quad (7.29)$$

where $A = \sum_{(p)} w_{ij}^2$, $B = \sum_{(p)} w_{ij}$, and $C = \sum_i (\sum_j w_{ij})^2$. The notation is now getting rather complicated, in particular the expression $\sum_i (\sum_j w_{ij})^2$. The calculation of this expression can best be thought of in terms of a number of steps:

(1) For each point add up the weights between it and all other points, to give $\sum_j w_{ij}$ for each point (each value of i).

(2) Square this total, to give $(\sum_j w_{ij})^2$ for each point.

(3) Add up all these squared totals, to give $\sum_i (\sum_j w_{ij})^2$.

Table 7.11 gives all the steps in the application of this form of the significance test to the situation shown in Figure 7.3.

(b) *Randomization*

$$E_I = -\frac{1}{n-1}$$

$$\sigma_I = \sqrt{\frac{\begin{array}{c} n[(n^2 + 3 - 3n)A + 3B^2 - nC] \\ - k[(n^2 - n)A + 6B^2 - 2nC] \end{array}}{(n-1)(n-2)(n-3)B^2}} \quad (7.30)$$

where, in addition to the notation explained above, k is the kurtosis of the point values. All the necessary calculations for the application of this form of significance test to the situation shown in Figure 7.3 are given in Table 7.11.

The underlying logic of the two forms of null hypothesis has already been discussed in connection with the significance test applied to areal data. With point data the same care must be taken in choosing the form of null hypothesis. In summary, the randomization null hypothesis only takes into account the particular set of points within the study region. The null hypothesis of normality involves the assumption that the point values within the study region can be regarded as a random sample of values

Table 7.11 Spatial Autocorrelation Measures for Point Data

(a) *Calculations for All Points*:

Point	x	$(x-\bar{x})$	$(x-\bar{x})^2$	$(x-\bar{x})^4$
A	8	−9	81	6561
B	5	−12	144	20736
C	10	−7	49	2401
D	42	25	625	390625
E	20	3	9	81
n = 5	85		908	420404

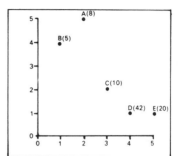

$$\bar{x} = \frac{\sum x}{n} = \frac{85}{5} = 17$$

$$\sigma = \sqrt{\frac{\sum(x-\bar{x})^2}{n}} = \sqrt{\frac{908}{5}} = 13.4759$$

$$k = \frac{\sum(x-\bar{x})^4}{n\sigma^4} = \frac{420404}{5 \times 13.4759^4} = 2.4596$$

(b) *Calculations for All Pairs of Points*:

Points i	j	x_i	x_j	$(x_i-\bar{x})$	$(x_j-\bar{x})$	Distance d_{ij}	Weight w_{ij}	w_{ij}^2	$w_{ij}(x_i-\bar{x})(x_j-\bar{x})$
A	B	8	5	−9	−12	1.4142	0.7071	0.5000	76.3675
A	C	8	10	−9	−7	3.1623	0.3162	0.1000	19.9224
A	D	8	42	−9	25	4.4721	0.2236	0.0500	−50.3115
A	E	8	20	−9	3	5.0000	0.2000	0.0400	−5.4000
B	C	5	10	−12	7	2.8284	0.3536	0.1250	29.6985
B	D	5	42	−12	25	4.2427	0.2357	0.0556	−70.7107
B	E	5	20	−12	3	5.0000	0.2000	0.0400	−7.2000
C	D	10	42	−7	25	1.4142	0.7071	0.5000	−123.7437
C	E	10	20	−7	3	2.2361	0.4472	0.2000	−9.3915
D	E	42	20	25	3	1.0000	1.0000	1.0000	75.0000
						4.3905	2.6106		−65.7690

c) *Calculations Relating to the Matrix of Weights*:

Point i	Point j A	B	C	D	E	$\sum_j w_{ij}$	$\left(\sum_j w_{ij}\right)^2$
A		0.7071	0.3162	0.2236	0.2000	1.4470	2.0936
B	0.7071		0.3536	0.2357	0.2000	1.4964	2.2391
C·	0.3162	0.3536		0.7071	0.4472	1.8241	3.3274
D	0.2236	0.2357	0.7071		1.0000	2.1664	4.6934
E	0.2000	0.2000	0.4472	1.0000		1.8472	3.4122
							15.7657

Table 7.11 continued

All the elements of the various equations have now been calculated:

$$A = \sum_{(p)} w_{ij}^2 = 2.6106$$

$$B = \sum_{(p)} w_{ij} = 4.3905$$

$$C = \sum_i \left(\sum_j w_{ij} \right)^2 = 15.7657$$

(d) *Calculation of I*:

$$I = \frac{n \sum_{(p)} w_{ij}(x_i - \bar{x})(x_j - \bar{x})}{(\sum_{(p)} w_{ij}) \sum (x - \bar{x})^2} = \frac{5 \times -65.7690}{4.3905 \times 908} = -0.0825$$

(e) *Normality Significance Test*:

$$E_I = -\frac{1}{n-1} = -\frac{1}{5-1} = -0.25$$

$$\sigma_I = \sqrt{\frac{n^2 A + 3B^2 - nC}{(n^2 - 1)B^2}} = \sqrt{\frac{(5^2 \times 2.6106) + (3 \times 4.3905^2) - (5 \times 15.7657)}{(5^2 - 1)4.3905^2}} = 0.3093$$

$$z = \frac{I - E_I}{\sigma_I} = \frac{-0.0825 - (-0.25)}{0.3093} = 0.5416$$

Assuming a two-tailed test at the 0.05 significance level, the observed degree of spatial autocorrelation is not significant (appropriate two-tailed critical value of $z = 1.960$).

(f) *Ramdomization Significance Test*:

$$E_I = -\frac{1}{n-1} = -\frac{1}{5-1} = -0.25$$

$$\sigma_I = \sqrt{\frac{n[(n^2 + 3 - 3n)A + 3B^2 - nC] - k[(n^2 - n)A + 6B^2 - 2nC]}{(n-1)(n-2)(n-3)B^2}}$$

$$= \sqrt{\frac{5[(5^2 + 3 - 3 \times 5)2.6016 + (3 \times 4.3905^2) - (5 \times 15.7657)] - 2.4596[(5^2 - 5)2.6106 + (6 \times 4.3905^2) - (2 \times 5 \times 15.7657)]}{(5-1)(5-2)(5-3)4.3905^2}}$$

$$= 0.2891$$

$$z = \frac{I - E_I}{\sigma_I} = \frac{-0.0825 - (-0.25)}{0.2891} = 0.5795$$

Assuming a two-tailed test at the 0.05 significance level, the observed degree of spatial autocorrelation is not significant (appropriate two-tailed critical value of $z = 1.960$).

drawn from a normally distributed population. In general randomization is the 'safer' choice, since it involves fewer assumptions about the data.

So far little has been said about the weights used. Each weight is intended to be a measure of the influence exerted by one point on another. The use of the reciprocal of distance ($1/d_{ij}$) as a weight implies that the influence decreases with distance. It could be argued, by analogy with gravitation, that influence is likely to decrease with the square of distance. In this case a suitable weight might be $1/d_{ij}^2$. Any other weight could be applied, provided it means something to the researcher and provided it can be justified theoretically. There is a danger, however, that one might be tempted to choose a set of weights which gives a 'satisfactory' result. This sort of approach is **not** scientific and should be studiously avoided.

7.6 Computer Programs

Program 12 — Mean Centre and Standard Distance

This program calculates the mean centre and standard distance for a set of points located by their X, Y coordinates. Both weighted and unweighted point data can be handled.

Program Listing

```
10 REM - MEAN CENTRE AND STANDARD DISTANCE
20 PRINT"-------------------------------"
30 PRINT"MEAN CENTRE AND STANDARD DISTANCE"
40 PRINT"-------------------------------"
50 PRINT
60 INPUT"Number of points",NX
70 NY=NX+NX
80 N=NX+NY
90 DIM X(N)
100 PRINT
110 PRINT"Now type in X coordinates (Eastings):"
120 PRINT"(Mistakes can be corrected later)"
130 FOR I=1 TO NX
140 PRINT"Point ";I;" ";
150 INPUT X(I)
160 NEXT I
170 PRINT
180 PRINT"Now type in Y coordinates (Northings):"
190 PRINT"(Mistakes can be corrected later)"
200 FOR I=NX+1 TO NY
210 PRINT"Point ";I-NX;" ";
220 INPUT X(I)
230 NEXT I
240 PRINT
250 INPUT"Are points to be weighted (Y/N)",W$
260 IF W$="Y" GOTO 320
270 IF W$<>"N" GOTO 250
280 FOR I=NY+1 TO N
290 X(I)=1
300 NEXT I
310 GOTO 390
320 PRINT
330 PRINT"Now type in point weights:"
340 PRINT"(Mistakes can be corrected later)"
350 FOR I=NY+1 TO N
360 PRINT"Point ";I-NY;" ";
370 INPUT X(I)
380 NEXT I
390 N1=1
400 N2=NX
410 V$="X"
420 GOSUB 810
430 N1=NX+1
440 N2=NY
450 V$="Y"
460 GOSUB 810
470 IF W$="N" GOTO 520
480 N1=NY+1
490 N2=N
500 V$="weight"
510 GOSUB 810
520 XW=0
530 YW=0
540 XX=0
550 YY=0
560 W=0
570 FOR I=1 TO NX
580 XW=XW+X(I)*X(I+NY)
590 YW=YW+X(I+NX)*X(I+NY)
600 XX=XX+X(I)^2*X(I+NY)
610 YY=YY+X(I+NX)^2*X(I+NY)
620 W=W+X(I+NY)
630 NEXT I
640 XW=XW/W
650 YW=YW/W
660 SD=SQR((XX/W-XW^2)+(YY/W-YW^2))
670 PRINT
680  PRINT"-------------------------------"
690 PRINT"MEAN CENTRE AND STANDARD DISTANCE"
700 PRINT"-------------------------------"
710 PRINT
720 IF W$="N" GOTO 750
730 PRINT"POINTS ARE WEIGHTED"
740 PRINT
750 PRINT"Coordinates of mean centre:"
760 PRINT"X = ";XW
770 PRINT"Y = ";YW
780 PRINT
790 PRINT"Standard distance = ";SD
800 STOP
810 REM - DATA CHECKING ROUTINE
820 PRINT
830 PRINT"Do you want to check/correct ";V$;
840 INPUT" values (Y/N)";Q$
850 IF Q$="N" GOTO 960
860 IF Q$<>"Y" GOTO 830
870 K=0
880 PRINT"Press RETURN if value is correct";
890 PRINT" - otherwise type correct value"
900 FOR I=N1 TO N2
910 K=K+1
920 PRINT K;" ";X(I);
930 INPUT A$
940 IF A$<>"" X(I)=VAL(A$)
950 NEXT I
960 RETURN
```

Example Output 1

```
------------------------------------
MEAN CENTRE AND STANDARD DISTANCE
------------------------------------

Number of points?8

Now type in X coordinates (Eastings):
(Mistakes can be corrected later)
Point 1 ?1
Point 2 ?2
Point 3 ?2
Point 4 ?2
Point 5 ?3
Point 6 ?3
Point 7 ?3
Point 8 ?4

Now type in Y coordinates (Northings):
(Mistakes can be corrected later)
Point 1 ?2
Point 2 ?3
Point 3 ?2
Point 4 ?1
Point 5 ?4
Point 6 ?3
Point 7 ?1
Point 8 ?4

Are points to be weighted (Y/N)?N

Do you want to check/correct X values (Y/N)?N

Do you want to check/correct Y values (Y/N)?N

------------------------------------
MEAN CENTRE AND STANDARD DISTANCE
------------------------------------

Coordinates of mean centre:
X = 2.5
Y = 2.5

Standard distance = 1.41421356

STOP at line 800
```

The data used are from the worked example in the text (Table 7.3). In this case the points are not weighted.

Example Output 2

```
------------------------------------
MEAN CENTRE AND STANDARD DISTANCE
------------------------------------

Number of points?5

Now type in X coordinates (Eastings):
(Mistakes can be corrected later)
Point 1 ?2
Point 2 ?1
Point 3 ?3
Point 4 ?4
Point 5 ?5

Now type in Y coordinates (Northings):
(Mistakes can be corrected later)
Point 1 ?5
Point 2 ?4
Point 3 ?2
Point 4 ?1
Point 5 ?1

Are points to be weighted (Y/N)?Y

Now type in point weights:
(Mistakes can be corrected later)
Point 1 ?8
Point 2 ?5
Point 3 ?10
Point 4 ?42
Point 5 ?20

Do you want to check/correct X values (Y/N)?N

Do you want to check/correct Y values (Y/N)?N

Do you want to check/correct weight values (Y/N)?N

------------------------------------
MEAN CENTRE AND STANDARD DISTANCE
------------------------------------

POINTS ARE WEIGHTED

Coordinates of mean centre:
X = 3.75294118
Y = 1.67058824

Standard distance = 1.7032434

STOP at line 800
```

The data used are from the weighted mean centre example in the text (Figure 7.3).

Program 13 — Standard Deviational Ellipse

This program calculates the parameters of a standard deviational ellipse, applied to either weighted or unweighted point data. Points are located by their X, Y coordinates.

Program Listing

```
10 REM - STANDARD DEVIATIONAL ELLIPSE
20 PRINT"--------------------------"
30 PRINT"STANDARD DEVIATIONAL ELLIPSE"
40 PRINT"--------------------------"
50 PRINT
60 INPUT"Number of points",NX
70 NY=NX+NX
80 N=NX+NY
90 DIM X(N)
100 PRINT
110 PRINT"Now type in X coordinates (Eastings):"
120 PRINT"(Mistakes can be corrected later)"
130 FOR I=1 TO NX
140 PRINT"Point ";I;" ";
150 INPUT X(I)
160 NEXT I
170 PRINT
180 PRINT"Now type in Y coordinates (Northings):"
190 PRINT"(Mistakes can be corrected later)"
200 FOR I=NX+1 TO NY
210 PRINT"Point ";I-NX;" ";
220 INPUT X(I)
230 NEXT I
240 PRINT
250 INPUT"Are points to be weighted (Y/N)",W$
260 IF W$="Y" GOTO 320
270 IF W$<>"N" GOTO 250
280 FOR I=NY+1 TO N
290 X(I)=1
300 NEXT I
310 GOTO 390
320 PRINT
330 PRINT"Now type in point weights:"
340 PRINT"(mistakes can be corrected later)"
350 FOR I=NY+1 TO N
360 PRINT"Point ";I-NY;" ";
370 INPUT X(I)
380 NEXT I
390 N1=1
400 N2=NX
410 V$="X"
420 GOSUB 1060
430 N1=NX+1
440 N2=NY
450 V$="Y"
460 GOSUB 1060
470 IF W$="N" GOTO 520
480 N1=NY+1
490 N2=N
500 V$="weight"
510 GOSUB 1060
520 XM=0
530 YM=0
540 W=0
550 FOR I=1 TO NX
560 XM=XM+X(I)*X(I+NY)
570 YM=YM+X(I+NX)*X(I+NY)
580 W=W+X(I+NY)
590 NEXT I
600 XM=XM/W
610 YM=YM/W
620 XX=0
630 YY=0
640 XY=0
650 FOR I=1 TO NX
660 X(I)=X(I)-XM
670 X(I+NX)=X(I+NX)-YM
680 XX=XX+X(I)^2*X(I+NY)
690 YY=YY+X(I+NX)^2*X(I+NY)
700 XY=XY+X(I)*X(I+NX)*X(I+NY)
710 NEXT I
720 T=ATN((XX-YY+SQR((XX-YY)^2+4*XY^2))/(2*XY))
730 IF T<0 T=T+1.57079633
740 ST=SIN(T)
750 CT=COS(T)
760 SX=SQR((XX*CT^2-2*XY*ST*CT+YY*ST^2)/W)
770 SY=SQR((XX*ST^2+2*XY*ST*CT+YY*CT^2)/W)
780 PRINT
790 PRINT"--------------------------"
800 PRINT"STANDARD DEVIATIONAL ELLIPSE"
810 PRINT"--------------------------"
820 IF W$="N" GOTO 850
830 PRINT
840 PRINT"POINTS ARE WEIGHTED"
850 PRINT
860 PRINT"Coordinates of centre:"
870 PRINT"X = ";XM
880 PRINT"Y = ";YM
890 PRINT
900 PRINT"Angle of rotation of Y axis = ";T*57.29577;
910 PRINT" degrees"
920 PRINT"(clockwise from original Y axis)"
930 PRINT
940 PRINT"Standard deviation along new X axis = ";SX
950 PRINT"Standard deviation along new Y axis = ";SY
960 PRINT
970 PRINT"Length of X axis = ";2*SX
980 PRINT"Length of Y axis = ";2*SY
990 PRINT
1000 PRINT"Area of ellipse = ";3.14159265*SX*SY*4
1010 R=SX/SY
1020 IF R<1 R=1/R
1030 PRINT
1040 PRINT"Ratio of long axis to short axis = ";R
1050 STOP
1060 REM - DATA CHECKING ROUTINE
1070 PRINT
1080 PRINT"Do you want to check/correct ";V$;
1090 INPUT" values (Y/N)";Q$
1100 IF Q$="N" GOTO 1210
1110 IF Q$<>"Y" GOTO 1080
1120 K=0
1130 PRINT"Press RETURN if value is correct";
1140 PRINT" - otherwise type correct value"
1150 FOR I=N1 TO N2
1160 K=K+1
1170 PRINT K;" ";X(I);
1180 INPUT A$
1190 IF A$<>"" X(I)=VAL(A$)
1200 NEXT I
1210 RETURN
```

Example Output 1

```
----------------------------
STANDARD DEVIATIONAL ELLIPSE
----------------------------

Number of points?8

Now type in X coordinates (Eastings):
(Mistakes can be corrected later)
Point 1 ?1
Point 2 ?2
Point 3 ?2
Point 4 ?2
Point 5 ?3
Point 6 ?3
Point 7 ?3
Point 8 ?4

Now type in Y coordinates (Northings):
(Mistakes can be corrected later)
Point 1 ?2
Point 2 ?3
Point 3 ?2
Point 4 ?1
Point 5 ?4
Point 6 ?3
Point 7 ?1
Point 8 ?4

Are points to be weighted (Y/N)?N

Do you want to check/correct X values (Y/N)?N

Do you want to check/correct Y values (Y/N)?N

----------------------------
STANDARD DEVIATIONAL ELLIPSE
----------------------------

Coordinates of centre:
X = 2.5
Y = 2.5

Angle of rotation of Y axis = 31.7174744 degrees
(clockwise from original Y axis)

Standard deviation along new X axis = 0.664065513
Standard deviation along new Y axis = 1.24860602

Length of X axis = 1.32813103
Length of Y axis = 2.49721204

Area of ellipse = 10.4194841

Ratio of long axis to short axis = 1.88024524

STOP at line 1050
```

The data used are from one of the worked examples in the text (Table 7.4 and Figure 7.11). In this case the points are not weighted. Note that the program also calculates the area of the ellipse and its index of elongation (ratio of long axis to short axis).

Example Output 2

```
----------------------------
STANDARD DEVIATIONAL ELLIPSE
----------------------------

Number of points?5

Now type in X coordinates (Eastings):
(Mistakes can be corrected later)
Point 1 ?2
Point 2 ?1
Point 3 ?3
Point 4 ?4
Point 5 ?5

Now type in Y coordinates (Northings):
(Mistakes can be corrected later)
Point 1 ?5
Point 2 ?4
Point 3 ?2
Point 4 ?1
Point 5 ?1

Are points to be weighted (Y/N)?Y

Now type in point weights:
(mistakes can be corrected later)
Point 1 ?8
Point 2 ?5
Point 3 ?10
Point 4 ?42
Point 5 ?20

Do you want to check/correct X values (Y/N)?N

Do you want to check/correct Y values (Y/N)?N

Do you want to check/correct weight values (Y/N)?N

----------------------------
STANDARD DEVIATIONAL ELLIPSE
----------------------------

POINTS ARE WEIGHTED

Coordinates of centre:
X = 3.75294118
Y = 1.67058824

Angle of rotation of Y axis = 50.7858116 degrees
(clockwise from original Y axis)

Standard deviation along new X axis = 1.64634529
Standard deviation along new Y axis = 0.43656072

Length of X axis = 3.29269057
Length of Y axis = 0.87312144

Area of ellipse = 9.03182356

Ratio of long axis to short axis = 3.77117136

STOP at line 1050
```

The weighted point data used here are from the weighted mean centre example in the text (Figure 7.3).

Program 14 — Nearest Neighbour Analysis

This program performs a nearest neighbour analysis on a set of points located by their X, Y coordinates.

Program Listing

```
10 REM - NEAREST NEIGHBOUR ANALYSIS
20 PRINT"------------------------"
30 PRINT"NEAREST NEIGHBOUR ANALYSIS"
40 PRINT"------------------------"
50 PRINT
60 INPUT"Number of points",NX
70 N=NX+NX
80 DIM X(N)
90 PRINT
100 PRINT"Now type in X coordinates (Eastings) of points:"
110 PRINT"(Mistakes can be corrected later)"
120 FOR I=1 TO NX
130 PRINT"Point ";I;" ";
140 INPUT X(I)
150 NEXT I
160 PRINT
170 PRINT"Now type in Y coordinates (Northings) of points:"
180 PRINT"(Mistakes can be corrected later)"
190 FOR I=NX+1 TO N
200 PRINT"Point ";I-NX;" ";
210 INPUT X(I)
220 NEXT I
230 N1=1
240 N2=NX
250 V$="X"
260 GOSUB 910
270 N1=NX+1
280 N2=N
290 V$="Y"
300 GOSUB 910
310 PRINT
320 PRINT"Now give the area of the study region:"
330 PRINT"If you want the analysis to be based only on"
340 PRINT"the enclosing rectangle, just press RETURN"
350 INPUT AR
360 X1=X(1)
370 X2=X(1)
380 Y1=X(1+NX)
390 Y2=X(1+NX)
400 FOR I=1 TO NX
410 IF X(I)<X1 X1=X(I)
420 IF X(I)>X2 X2=X(I)
430 IF X(I+NX)<Y1 Y1=X(I+NX)
440 IF X(I+NX)>Y2 Y2=X(I+NX)
450 XM=XM+X(I)
460 YM=YM+X(I+NX)
470 NEXT I
480 A=(X2-X1)*(Y2-Y1)
490 DD=(X2-X1)^2+(Y2-Y1)^2+1
500 XM=XM/NX
510 YM=YM/NX
520 ND=0
530 FOR I=1 TO NX
540 DM=DD
550 FOR J=1 TO NX
560 IF I=J GOTO 600
570 D=(X(I)-X(J))^2+(X(I+NX)-X(J+NX))^2
580 IF D>DM GOTO 600
590 DM=D
600 NEXT J
610 IF DM=0 GOTO 630
620 ND=ND+SQR(DM)
630 NEXT I
640 PRINT
650 PRINT"------------------------"
660 PRINT"NEAREST NEIGHBOUR ANALYSIS"
670 PRINT"------------------------"
680 PRINT
690 ND=ND/NX
700 P=NX/A
710 DR=0.5/SQR(P)
720 R=ND/DR
730 SE=0.26136/SQR(NX*P)
740 C=(ND-DR)/SE
750 PRINT"Based on enclosing rectangle:"
760 PRINT"Area = ";A
770 PRINT"Nearest neighbour index = ";R
780 PRINT"c = ";C
790 IF AR=0 GOTO 900
800 P=NX/AR
810 DR=0.5/SQR(P)
820 R=ND/DR
830 SE=0.26136/SQR(NX*P)
840 C=(ND-DR)/SE
850 PRINT
860 PRINT"Based on area as input:"
870 PRINT"Area = ";AR
880 PRINT"Nearest neighbour index = ";R
890 PRINT"c = ";C
900 STOP
910 REM - DATA CHECKING ROUTINE
920 PRINT
930 PRINT"Do you want to check/correct ";V$;
940 INPUT" values (Y/N)";Q$
950 IF Q$="N" GOTO 1060
960 IF Q$<>"Y" GOTO 930
970 K=0
980 PRINT"Press RETURN if value is correct";
990 PRINT" - otherwise type correct value"
1000 FOR I=N1 TO N2
1010 K=K+1
1020 PRINT K;" ";X(I);
1030 INPUT A$
1040 IF A$<>"" X(I)=VAL(A$)
1050 NEXT I
1060 RETURN
```

Example Output

```
----------------------------
NEAREST NEIGHBOUR ANALYSIS
----------------------------

Number of points?10

Now type in X coordinates (Eastings) of points:
(Mistakes can be corrected later)
Point 1 ?5
Point 2 ?2
Point 3 ?3
Point 4 ?1
Point 5 ?4
Point 6 ?2
Point 7 ?8
Point 8 ?7
Point 9 ?9
Point 10 ?6

Now type in Y coordinates (Northings) of points:
(Mistakes can be corrected later)
Point 1 ?9
Point 2 ?8
Point 3 ?8
Point 4 ?7
Point 5 ?7
Point 6 ?6
Point 7 ?4
Point 8 ?2
Point 9 ?2
Point 10 ?1

Do you want to check/correct X values (Y/N)?N

Do you want to check/correct Y values (Y/N)?N

Now give the area of the study region:
If you want the analysis to be based only on
the enclosing rectangle, just press RETURN
?55

----------------------------
NEAREST NEIGHBOUR ANALYSIS
----------------------------

Based on enclosing rectangle:
Area = 64
Nearest neighbour index = 1.22879815
c = 1.38415075

Based on area as input:
Area = 55
Nearest neighbour index = 1.32552887
c = 1.96933861

STOP at line 900
```

The data used are from Exercise 7.6 (Figure 7.19). Note the difference in the nearest neighbour index produced by changing the definition of the study area.

Program 15 — Join Count Statistics

The program calculates join count statistics for two-category ('black'/'white') areal data. Both free-sampling and non-free sampling null hypotheses can be tested. Various tests on the validity of the data are applied, and the user is notifed of obvious errors. For example, the input data are checked to see whether the total number of joins corresponds to the join data typed in for each area.

Program Listing

```
10 REM - JOIN COUNTS FOR NOMINAL DATA
20 PRINT"--------------------------"
30 PRINT"JOIN COUNTS FOR NOMINAL DATA"
40 PRINT"--------------------------"
50 PRINT
60 INPUT"Number of areas",N
70 PRINT
80 INPUT"Number of 'black' areas",B
90 PRINT
100 INPUT"Number of 'white' areas",W
110 IF B+W=N GOTO 140
120 PRINT"Impossible - please check your data!"
130 GOTO 60
140 DIM X(N)
150 PRINT
160 INPUT"Total number of joins",J
170 PRINT
180 PRINT"Now type in number of joins for each area:"
190 PRINT"(mistakes can be corrected later)"
200 FOR I=1 TO N
210 PRINT"Area ";I;" ";
220 INPUT X(I)
230 NEXT I
240 N1=1
250 N2=N
260 V$="join"
270 GOSUB 740
280 L=0
290 LJ=0
300 FOR I=1 TO N
310 L=L+X(I)*(X(I)-1)
320 LJ=LJ+X(I)
330 NEXT I
340 IF J=LJ/2 GOTO 390
350 PRINT"ERROR - join data incompatible with total";
360 PRINT" number of joins"
370 PRINT"Please check your data and try again"
380 STOP
390 PRINT
400 INPUT"Number of 'black'/'white' joins",BW
410 PRINT
420 PRINT"Probability that an area will be 'black'"
430 PRINT"(for free sampling null hypothesis)"
440 INPUT"If this is not applicable, just press RETURN",P
450 IF P=0 GOTO 610
```

```
460 IF P>0 AND P<1 GOTO 490
470 PRINT"Probability must be between 0 and 1 - try aga
480 GOTO 420
490 Q=1-P
500 E=2*J*P*Q
510 S=SQR((2*J+L)*P*Q-4*(J+L)*P^2*Q^2)
520 Z=(BW-E)/S
530 PRINT
540 PRINT"--------------------------"
550 PRINT"JOIN COUNTS FOR NOMINAL DATA"
560 PRINT"--------------------------"
570 PRINT
580 PRINT"Free sampling null hypothesis:"
590 PRINT"Expected number of black/white joins = ";E
600 PRINT"z = ";Z
610 N1=N*(N-1)
620 E=2*J*B*W/N1
630 N3=N*(N-1)*(N-2)*(N-3)
640 B1=B*(B-1)
650 W1=W*(W-1)
660 J1=J*(J-1)-L
670 S=SQR(E+L*B*W/N1+4*J1*B1*W1/N3-E^2)
680 Z=(BW-E)/S
690 PRINT
700 PRINT"Non-free sampling null hypothesis:"
710 PRINT"Expected number of black/white joins = ";E
720 PRINT"z = ";Z
730 STOP
740 REM - DATA CHECKING ROUTINE
750 PRINT
760 PRINT"Do you want to check/correct ";V$;
770 INPUT" values (Y/N)";Q$
780 IF Q$="N" GOTO 890
790 IF Q$<>"Y" GOTO 760
800 K=0
810 PRINT"Press RETURN if value is correct";
820 PRINT" - otherwise type correct value"
830 FOR I=N1 TO N2
840 K=K+1
850 PRINT K;" ";X(I);
860 INPUT A$
870 IF A$<>"" X(I)=VAL(A$)
880 NEXT I
890 RETURN
```

Example Output

```
-----------------------------
JOIN COUNTS FOR NOMINAL DATA
-----------------------------

Number of areas?11

Number of 'black' areas?5

Number of 'white' areas?6

Total number of joins?19

Now type in number of joins for each area:
(mistakes can be corrected later)
Area 1 ?2
Area 2 ?5
Area 3 ?4
Area 4 ?2
Area 5 ?3
Area 6 ?4
Area 7 ?6
Area 8 ?4
Area 9 ?3
Area 10 ?4
Area 11 ?1

Do you want to check/correct join values (Y/N)?N

Number of 'black'/'white' joins?6

Probability that an area will be 'black'
(for free sampling null hypothesis)
If this is not applicable, just press RETURN?0.4


-----------------------------
JOIN COUNTS FOR NOMINAL DATA
-----------------------------

Free sampling null hypothesis:
Expected number of black/white joins = 9.12
z = -1.291419

Non-free sampling null hypothesis:
Expected number of black/white joins = 10.3636364
z = -2.462348

STOP at line 730
```

The data used are from the worked example in the text (Figure 7.21a).

Program 16 — Spatial Autocorrelation for Areal Data

This program calculates Moran's coefficient of spatial autocorrelation for interval areal data. Both the normality and randomization significance tests are applied.

Program Listing

```
10 REM - SPATIAL AUTOCORRELATION FOR AREAL DATA
20 PRINT"------------------------------------"
30 PRINT"SPATIAL AUTOCORRELATION FOR AREAL DATA"
40 PRINT"------------------------------------"
50 PRINT
60 INPUT"Number of areas",N
70 INPUT"Total number of joins",J
80 NT=N+J+J
90 DIM X(NT),L(J)
100 PRINT
110 PRINT"Now type in values for each area:"
120 PRINT"(mistakes can be corrected later)"
130 FOR I=1 TO N
140 PRINT"Area ";I;" ";
150 INPUT X(I)
160 NEXT I
170 PRINT
180 PRINT"Now type in data for each join:"
190 PRINT"(mistakes can be corrected later)"
200 K=0
210 FOR I=N+1 TO N+J
220 K=K+1
230 PRINT"Join ";K;" separates area ";
240 INPUT X(I)
250 IF X(I)>0 AND X(I)<=J GOTO 280
260 PRINT"No such area - try again!"
270 GOTO 230
280 INPUT"from area ",X(I+J)
290 IF X(I+J)>0 AND X(I+J)<=J GOTO 320
300 PRINT"No such area - try again!"
310 GOTO 280
320 NEXT I
330 N1=1
340 N2=N
350 V$="area"
360 GOSUB 840
370 GOSUB 1000
380 XM=0
390 FOR I=1 TO N
400 XM=XM+X(I)
410 NEXT
420 XM=XM/N
430 XX=0
440 X4=0
450 FOR I=1 TO N
460 X(I)=X(I)-XM
470 XX=XX+X(I)^2
480 X4=X4+X(I)^4
490 NEXT I
500 SX=XX/N
510 K=X4/(N*SX^2)
520 XY=0
530 FOR I=N+1 TO N+J
540 XY=XY+X(X(I))*X(X(I+J))
550 L(X(I))=L(X(I))+1
560 L(X(I+J))=L(X(I+J))+1
570 NEXT I
580 L2=0
590 FOR I=1 TO J
600 L2=L2+L(I)^2
610 NEXT I
620 MI=(N*XY)/(J*XX)
630 E=-1/(N-1)
640 N2=N^2
650 J2=J^2
660 S=SQR((N2*J+3*J2-N*L2)/(J2*(N2-1)))
670 Z=(MI-E)/S
680 PRINT
690 PRINT"------------------------------------"
700 PRINT"SPATIAL AUTOCORRELATION FOR AREAL DATA"
710 PRINT"------------------------------------"
720 PRINT
730 PRINT"I = ";MI
740 PRINT"Expected value of I = ";E
750 PRINT"Normality significance test:"
760 PRINT"z = ";Z
770 S=N*(J*(N2+3-3*N)+3*J2-N*L2)
780 S=S-K*(J*(N2-N)+6*J2-2*N*L2)
790 S=SQR(S/(J2*(N-1)*(N-2)*(N-3)))
800 Z=(MI-E)/S
810 PRINT"Randomization significance test:"
820 PRINT"z = ";Z
830 STOP
840 REM - DATA CHECKING ROUTINE
850 PRINT
860 PRINT"Do you want to check/correct ";V$;
870 INPUT" values (Y/N)";Q$
880 IF Q$="N" GOTO 990
890 IF Q$<>"Y" GOTO 860
900 K=0
910 PRINT"Press RETURN if value is correct";
920 PRINT" - otherwise type correct value"
930 FOR I=N1 TO N2
940 K=K+1
950 PRINT K;" ";X(I);
960 INPUT A$
970 IF A$<>"" X(I)=VAL(A$)
980 NEXT I
990 RETURN
1000 REM - DATA CHECKING FOR JOIN DATA
1010 PRINT
1020 PRINT"Do you want to check/correct join data";
1030 INPUT" (Y/N)",Q$
1040 IF Q$="N" GOTO 1190
1050 IF Q$<>"Y" GOTO 1020
1060 K=0
1070 PRINT"Press RETURN if value is correct";
1080 PRINT" - otherwise type correct value"
1090 FOR I=N+1 TO N+J
1100 K=K+1
1110 PRINT"Join ";K
1120 PRINT"separates area ";X(I);
1130 INPUT A$
1140 IF A$<>"" X(I)=VAL(A$)
1150 PRINT"from area ";X(I+J);
1160 INPUT A$
1170 IF A$<>"" X(I+J)=VAL(A$)
1180 NEXT I
1190 RETURN
```

Example Output

```
----------------------------------------
SPATIAL AUTOCORRELATION FOR AREAL DATA
----------------------------------------


Number of areas?6
Total number of joins?9

Now type in values for each area:
(mistakes can be corrected later)
Area 1 ?2.6
Area 2 ?0.5
Area 3 ?2.4
Area 4 ?0.3
Area 5 ?3.8
Area 6 ?0.6

Now type in data for each join:
(mistakes can be corrected later)
Join 1 separates area ?1
from area ?2
Join 2 separates area ?1
from area ?3
Join 3 separates area ?2
from area ?3
Join 4 separates area ?2
from area ?4
Join 5 separates area ?3
from area ?4
Join 6 separates area ?3
from area ?5
Join 7 separates area ?4
from area ?5
Join 8 separates area ?5
from area ?6
Join 9 separates area ?4
from area ?6

Do you want to check/correct area values (Y/N)?N

Do you want to check/correct join data (Y/N)?N

----------------------------------------
SPATIAL AUTOCORRELATION FOR AREAL DATA
----------------------------------------

I = -0.182817
Expected value of I = -0.2
Normality significance test:
z = 0.061825
Randomization significance test:
z = 0.059140

STOP at line 830
```

The data used are from the worked example in
the text (Figure 7.23).

Program 17 — Spatial Autocorrelation for Point Data

This program calculates Moran's coefficient of spatial autocorrelation for a set of point values located by their X, Y coordinates. Both the normality and randomization significance tests are applied.

Program Listing

```
10 REM - SPATIAL AUTOCORRELATION FOR POINT DATA
20 PRINT"---------------------------------------"
30 PRINT"SPATIAL AUTOCORRELATION FOR POINT DATA"
40 PRINT"---------------------------------------"
50 PRINT
60 INPUT"Number of points",N
70 NN=N+N
80 N3=NN+N
90 DIM X(N3)
100 PRINT
110 PRINT"Now type in X coordinates (Eastings) of points:"
120 PRINT"(mistakes can be corrected later)"
130 FOR I=1 TO N
140 PRINT"Point ";I;" ";
150 INPUT X(I)
160 NEXT I
170 PRINT
180 PRINT"Now type in Y coordinates (Northings) of points:"
190 PRINT"(mistakes can be corrected later)"
200 FOR I=N+1 TO NN
210 PRINT"Point ";I-N;" ";
220 INPUT X(I)
230 NEXT I
240 PRINT
250 PRINT"Now type in point values:"
260 PRINT"(mistakes can be corrected later)"
270 FOR I=NN+1 TO N3
280 PRINT"Point ";I-NN;
290 INPUT X(I)
300 NEXT I
310 N1=1
320 N2=NN
330 V$="X coordinate"
340 GOSUB 970
350 N1=N+1
360 N2=NN
370 V$="Y coordinate"
380 GOSUB 970
390 N1=NN+1
400 N2=N3
410 V$="point"
420 GOSUB 970
430 XM=0
440 FOR I=1 TO N
450 XM=XM+X(I+NN)
460 NEXT I
470 XM=XM/N
480 FOR I=NN+1 TO N3
490 X(I)=X(I)-XM
500 NEXT I
510 X2=0
520 X4=0
530 FOR I=1 TO N
540 W2=0
550 IY=I+N
560 ID=IY+N
```

```
570 X2=X2+X(ID)^2
580 X4=X4+X(ID)^4
590 FOR J=1 TO N
600 IF I=J GOTO 690
610 JY=J+N
620 JD=JY+N
630 W=(X(I)-X(J))^2+(X(IY)-X(JY))^2
640 IF I<J A=A+1/W
650 W=1/SQR(W)
660 IF I<J B=B+W
670 IF I<J WX=WX+W*X(ID)*X(JD)
680 W2=W2+W
690 NEXT J
700 C=C+W2^2
710 NEXT I
720 SX=X2/N
730 K=X4/(N*SX^2)
740 MI=(N*WX)/(B*X2)
750 E=-1/(N-1)
760 B2=B^2
770 N2=N^2
780 S=N2*A+3*B2-N*C
790 S=SQR(S/((N2-1)*B2))
800 Z=(MI-E)/S
810 PRINT
820 PRINT"---------------------------------------"
830 PRINT"SPATIAL AUTOCORRELATION FOR POINT DATA"
840 PRINT"---------------------------------------"
850 PRINT
860 PRINT"I = ";MI
870 PRINT"Expected value of I = ";E
880 PRINT"Normality significance test:"
890 PRINT"z = ";Z
900 S=N*(A*(N2+3-3*N)+3*B2-N*C)
910 S=S-K*(A*(N2-N)+6*B2-2*N*C)
920 S=S/(B2*(N-1)*(N-2)*(N-3))
930 Z=(MI-E)/SQR(S)
940 PRINT"Randomization significance test:"
950 PRINT"z = ";Z
960 STOP
970 REM - DATA CHECKING ROUTINE
980 PRINT
990 PRINT"Do you want to check/correct ";V$;
1000 INPUT" values (Y/N)";Q$
1010 IF Q$="N" GOTO 1120
1020 IF Q$<>"Y" GOTO 990
1030 K=0
1040 PRINT"Press RETURN if value is correct";
1050 PRINT" - otherwise type correct value"
1060 FOR I=N1 TO N2
1070 K=K+1
1080 PRINT K;" ";X(I);
1090 INPUT A$
1100 IF A$<>"" X(I)=VAL(A$)
1110 NEXT I
1120 RETURN
```

Example Output

```
------------------------------------------
SPATIAL AUTOCORRELATION FOR POINT DATA
------------------------------------------

Number of points?5

Now type in X coordinates (Eastings) of points:
(mistakes can be corrected later)
Point 1 ?2
Point 2 ?1
Point 3 ?3
Point 4 ?4
Point 5 ?5

Now type in Y coordinates (Northings) of points:
(mistakes can be corrected later)
Point 1 ?5
Point 2 ?4
Point 3 ?2
Point 4 ?1
Point 5 ?1

Now type in point values:
(mistakes can be corrected later)
Point 1?8
Point 2?5
Point 3?10
Point 4?42
Point 5?20

Do you want to check/correct X coordinate values (Y/N)?N

Do you want to check/correct Y coordinate values (Y/N)?N

Do you want to check/correct point values (Y/N)?N

------------------------------------------
SPATIAL AUTOCORRELATION FOR POINT DATA
------------------------------------------

I = -0.082488
Expected value of I =-0.25
Normality significance test:
z = 0.541545
Randomization significance test:
z = 0.579537

STOP at line 960
```

The data used are from the worked example in
the text (Table 7.11).

References

Boyce, R. R. and Clark, W. A. V., 1964, The concept of shape in geography, *Geographical Review, 54,* 561–572

Bunge, W., 1966, *Theoretical Geography,* London

Clark, P. J. and Evans, F. C., 1954, Distance to nearest neighbor as a measure of spatial relationships in populations, *Ecology, 35,* 445–453

Cliff, A. D., Martin, R. L. and Ord, J. K., 1975, A test of spatial autocorrelation in choropleth maps based upon a modified χ^2 statistic, *Transactions of the Institute of British Geographers, 65,* 109–129

Cliff, A. D. and Ord, J. K., 1973, *Spatial Autocorrelation,* London

Cole, J. P., 1975, *Situations in Human Geography: a Practical Approach,* Oxford

Cole, J. P. and King, C. A. M., 1968, *Quantitative Geography,* London

Dacey, M. F., 1968, A review on measures of contiguity for two and k-colour maps, in: Berry, B. J. L. and Marble, D. F. (eds), *Spatial Analysis: a Reader in Statistical Geography,* Englewood Cliffs

Dawson, A. H., 1975, Are geographers engaging in a landscape lottery?, *Area, 7,* 42–45

Ebdon, D. S., 1976, On the underestimation inherent in the commonly-used formulae, *Area, 8,* 165–169.

Hammond, R. and McCullagh, P. S., 1974, *Quantitative Techniques in Geography: an Introduction,* Oxford

Kershaw, K. A., 1964, *Quantitative and Dynamic Ecology,* London

King, L. J., 1969, *Statistical Analysis in Geography,* Englewood Cliffs

Krishna Iyer, P. V., 1949, The first and second moments of some probability distributions arising from points on a lattice and their applications, *Biometrika, 36,* 135–141.

Kruskal, W. H. and Wallis, W. A., 1952, Use of ranks in one-criterion variance analysis, *Journal of the American Statistical Association, 47,* 583–621

Lieberman, B. (ed), 1971, *Contemporary Problems in Statistics: a Book of Readings for the Behavioral Sciences,* New York

Mather, P. M., 1976, *Computers in Geography: a Practical Approach,* Oxford

Moran, P. A. P., 1950, Notes on continuous stochastic phenomena, *Biometrika, 37,* 17–23

Neft, D. S., 1966, *Statistical Analysis for Spatial Distributions,* Philadelphia (Regional Science Research Institute Monograph Series, No. 2)

Pounds, N. J., 1963, *Political Geography,* New York

Siegel, S., 1956, *Nonparametric Statistics for the behavioral Sciences,* New York

Taylor, P. J., 1971, Distances within shapes: an introduction to a family of finite frequency distributions, *Geografiska Annaler,* B, *53,* 40–53

APPENDIX A

Answers to Exercises

1.1

Class boundaries	Standard deviates	Normal probabilities
2	−2.363	
4	−1.562	0.051
6	−0.760	0.165
8	0.042	0.292
10	0.843	0.284
12	1.645	0.151
14	2.446	0.041
16	3.248	0.008

1.2

(a) 5 deaths: $\dfrac{z^5}{5!e^z} = \dfrac{0.61^5}{120 \times 1.8404}$

$$= 0.0004$$

(b) 6 deaths: $\dfrac{z^6}{6!e^z} = \dfrac{0.61^6}{720 \times 1.8404}$

$$= 0.00004$$

(c) 10 deaths: $\dfrac{z^{10}}{10!e^z} = \dfrac{0.61^{10}}{3628800 \times 1.8404}$

$$= 0.000000001$$

1.3 *See page 178*

2.1

(a) mode = 0.00, median = 0.00, mean = 7.626

(b) there is no single mode, median = 2.76, mean = 3.021

(c) mode = 6.9, median = 7.65, mean = 8.454

2.2

(a) range = 23.85 − 9.42 = 14.43, upper quartile = 14.14, lower quartile = 12.20, inter-quartile range = 14.14 − 12.20 = 1.94

(b) range = 14.4 − 2.0 = 12.4, upper quartile = 9.80, lower quartile = 6.50, inter-quartile range = 9.80 − 6.10 = 3.30 (n.b. there are only 49 values of this variable)

(c) range = 2.56 − 0.26 = 2.30, upper quartile = 0.79, lower quartile = 0.45, inter-quartile range = 0.79 − 0.45 = 0.34

2.3

standard deviation = 3.055

2.4

x	$(x-\bar{x})$	$(x-\bar{x})^2$	$(x-\bar{x})^3$	$(x-\bar{x})^4$
2	−1	1	−1	1
2	−1	1	−1	1
1	−2	4	−8	16
3	0	0	0	0
4	1	1	1	1
7	4	16	64	256
3	0	0	0	0
2	−1	1	−1	1
4	1	1	1	1
2	−1	1	−1	1
30		26	54	278

(a)

$\bar{x} = 3$ $\qquad \sigma = \sqrt{\dfrac{26}{10}} = 1.61245$

1.3

$$z = \frac{160}{400} = 0.4 \text{ (campers per square)}$$

Campers	Poisson term	Probability	Frequency
0	$\dfrac{1}{e^{0.4}} = \dfrac{1}{1.4918}$	0.6703	268 (0.6703×400)
1	$\dfrac{0.4}{e^{0.4}} = \dfrac{0.4}{1.4918}$	0.2681	107 (0.2681×400)
2	$\dfrac{0.4^2}{2!e^{0.4}} = \dfrac{0.16}{2 \times 1.4918}$	0.0536	21 (0.0536×400)
3	$\dfrac{0.4^3}{3!e^{0.4}} = \dfrac{0.064}{6 \times 1.4918}$	0.0072	3 (0.0072×400)
4	$\dfrac{0.4^4}{4!e^{0.4}} = \dfrac{0.0256}{24 \times 1.4918}$	0.0007	0 (0.0007×400)
5	$\dfrac{0.4^5}{5!e^{0.4}} = \dfrac{0.01024}{120 \times 1.4918}$	0.00006	0 (0.00006×400)
6	$\dfrac{0.4^6}{6!e^{0.4}} = \dfrac{0.004096}{720 \times 1.4918}$	0.000004	0 (0.000004×400)

2.4 *continued*

$$\text{skewness} = \frac{54}{10 \times 1.61245^3} = 1.288$$

$$\text{kurtosis} = \frac{278}{10 \times 1.61245^4} = 4.112$$

(b) $\bar{x} = 7.896$ $n = 49$

$$\sigma = \sqrt{\frac{\sum (x - \bar{x})^2}{n}} = \sqrt{\frac{298.859}{49}}$$
$$= 2.470$$

$$\text{skewness} = \frac{\sum (x - \bar{x})^3}{n\sigma^3} = \frac{160.792}{49 \times 2.470^3} = 0.218$$

$$\text{kurtosis} = \frac{\sum (x - \bar{x})^4}{n\sigma^4} = \frac{5593.640}{49 \times 2.470^4} = 3.068$$

3.1
Worked example (taking every fourth house from Table 3.1):
New Commonwealth Immigrants   3/50 = 6%

Television ownership   43/50 = 86%
Car ownership   23/50 = 46%
NDP voters   17/50 = 34%

3.2
Only method 6 fully satisfies the two criteria, and this method is only completely satisfactory if the addresses are returned to the hat after each draw (as explained later in Section 3.1).

3.3

$$\frac{1}{200} \times \frac{1}{200} \times \frac{1}{200} \cdots \text{ and so on, 50 times} \simeq$$
0.000000000003

3.4
See Figure A1.

3.5
(a) As soon as the required 5% sample of people has been selected from any particular

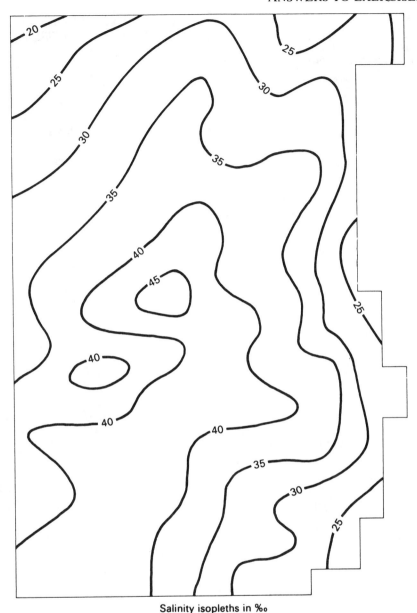

Salinity isopleths in ‰

Figure A1

ward, no more people can be selected from that ward. The remaining individuals in the ward no longer have the same probability of selection as individuals in any other ward in which the 5% quota has not yet been filled.

(b) Not all households are entered on the electoral register. Some may forget or refuse to fill in the necessary form. These households clearly have no chance of selection. More importantly, the electoral register contains one entry for each individual in the household who is entitled to vote, not one entry per household. Households which contain several voters will therefore stand a better chance (have a higher probability) of being selected than those containing only one or two voters.

3.6
- (a) 3.47 to 3.53 (3.5 ± 0.03)
- (b) 3.44 to 3.56 (3.5 ± 0.06)
- (c) 3.41 to 3.59 (3.5 ± 0.09)

3.7

$$SE_s = \frac{s}{\sqrt{2n}} = \frac{1.5}{\sqrt{4000}} = 0.0237$$

At the 0.997 level $\sigma = s \pm 3S_s = 1.5 \pm 3 \times 0.0237$

The population standard deviation can therefore be expected to lie between 1.4289 and 1.5711.

3.8

$$SE_s = \frac{\hat{\sigma}}{\sqrt{2n}} = \frac{1}{\sqrt{10}} = 0.3162$$

At the 0.954 level $\sigma = s \pm 2SE_s = 0.895 \pm 2 \times 0.3162$.

The population standard deviation can therefore be expected to lie between 0.263 and 1.527.

3.9

At the 0.954 level population % = sample % ± 2SE%

$$SE\% = \sqrt{\frac{pq}{n}} = \sqrt{\frac{46 \times 54}{50}} = 7.048$$

The 0.954 limits are therefore $46 \pm 2 \times 7.048 = 31.904$ and 60.096.

3.10

$\bar{x} = 3.5$ $s = 1.5$. To obtain an accuracy of ±0.5 at the 0.997 level $SE_{\bar{x}}$ needs to be $0.5/3 = 0.167$ (the 0.997 limits are set by $3SE_{\bar{x}}$).

$$n = \left(\frac{s}{SE_{\bar{x}}}\right)^2 = \left(\frac{1.5}{0.167}\right)^2 = 80.676$$

A sample size of at least 81 would therefore be necessary.

3.11

(a) Sample % = 34, i.e. p = 34 and q = 66 (100 − 34). To obtain an accuracy of ±15% at the 0.997 level, SE% needs to be 15/3 = 5 (the 0.997 limits are set by 3SE%).

$$n = \frac{pq}{(SE\%)^2} = \frac{34 \times 66}{5^2} = 89.76$$

A sample size of at least 90 would therefore be necessary.

(b) Results for discussion in class.

4.1 *See page 181*

4.2

b contains 10 runs and d contains 6 runs.

4.3

(a)

Sequence	Runs	Sequence	Runs
111000	2	011100	3
110100	4	011010	5
110010	4	011001	4
110001	3	010110	5
101100	4	010101	6
101010	6	010011	4
101001	5	001110	3
100110	4	001101	4
100101	5	001011	4
100011	3	000111	2

(b) 0.1 (2/20)
(c) 0.3 (6/20)
(d) Median = 17. The sequence of busy and slack periods is therefore:

SS BBBBBBB S B S BBB SSSSS B SSS

There are 9 runs, with $n_1 = 12$ and $n_2 = 12$. The critical values for the number of runs at the 0.05 level (from Appendix C2) are 7 and 19. Since the observed number of runs is not less than or equal to the lower critical value, or greater than or equal to the higher critical value, the null hypothesis can not be rejected. The observed sequence of busy and slack periods is not significantly different from random at the 0.05 level.

4.1

Class	Frequency	Probabilities observed	normal	Cumulative observed	normal	Absolute difference
2–3.99	3	0.0612	0.051	0.0612	0.051	0.0102
4–5.99	7	0.0142	0.165	0.2041	0.216	0.0119
6–7.99	18	0.3673	0.292	0.5714	0.508	0.0634
8–9.99	10	0.2041	0.284	0.7755	0.792	0.0165
10–11.99	9	0.1837	0.151	0.9592	0.943	0.0162
12–13.99	1	0.0204	0.041	0.9796	0.984	0.0044
14–15.99	1	0.0204	0.008	1.0000	0.992	0.0080
	49	1.0000	0.992			

D (maximum absolute difference) = 0.0634. At the 0.05 level the critical value of D with 49 degrees of freedom (from Appendix C1) is 0.19. Since the calculated value of D is less than the critical value the null hypothesis can not be rejected. The observed distribution is not significantly different from a normal distribution at the 0.05 level.

4.4

x	r_x	y	r_y
9.2	16	5.5	4.5
6.5	7	15.5	24
6.8	8	5.5	4.5
10.7	20	8.0	15
14.4	23	5.9	6
5.4	3	7.5	12
6.9	9	3.8	2
7.7	13.5	7.2	11
11.9	21	10.1	18
13.0	22	10.5	19
		2.0	1
		7.0	10
		10.0	17
		7.7	13.5
	142.5		157.5

$n_x = 10$ $n_y = 14$

$$U_x = 10 \times 14 + \frac{10(10 + 1)}{2} - 142.5 = 52.5$$

$$U_y = 10 \times 14 + \frac{14(14 + 1)}{2} - 157.5 = 87.5$$

Since the alternative hypothesis is directional (H_1: X > Y), a one-tailed test is appropriate. The relevant value of U for the test is U_x =

52.5. The critical value of U for a one-tailed test with $n_x = 10$ and $n_y = 14$ is 41 (from Appendix C3a). As the calculated value of U is greater than the critical value, the null hypothesis can not be rejected. The proportion of children in local authority care is not significantly greater (at the 0.05 level) in 'large' towns than in 'small' towns.

4.5

x	y	x^2	y^2
11.11	9.76	123.432	95.258
8.64	8.09	74.650	65.448
10.11	8.02	102.212	64.321
13.07	11.50	170.825	132.250
7.71	7.50	59.444	56.250
9.68	11.47	93.702	131.561
10.19	6.52	103.836	42.511
10.13	12.83	102.617	164.609
7.74	6.58	59.908	43.297
12.29	8.91	151.044	79.388
	9.50		90.250
	8.54		72.932
	7.34		53.876
	9.37		87.797
	6.30		39.690
100.67	132.23	1041.670	1219.435

$\bar{x} = 100.67/10 = 10.067$

$\bar{y} = 132.23/15 = 8.815$

$$t = \frac{|10.067 - 8.815|}{\sqrt{\dfrac{\dfrac{1041.670}{10} - 10.067^2}{9} + \dfrac{\dfrac{1219.435}{15} - 8.815^2}{14}}}$$

$= 1.658$

$$df = \frac{(0.3136 + 0.2565)^2}{0.3136^2/9 + 0.2565^2/14} = 20.8 \simeq 20$$

The critical value of t for a one-tailed test at the 0.01 level with 20 degrees of freedom (from Appendix C4) is 2.53. The mean of x is greater than the mean of y, so the difference is in the specified direction. However, since the calculated value of t is less than the critical value, the null hypothesis can not be rejected. Northern towns are not significantly wetter than southern towns at the 0.01 level.

4.6

(a) $A = \dfrac{560 \times 100}{1450} = 38.6$

$B = \dfrac{320 \times 100}{1450} = 22.1$ $C = \dfrac{180 \times 100}{1450} = 12.4$

$D = \dfrac{150 \times 100}{1450} = 10.3$ $E = \dfrac{240 \times 100}{1450} = 16.5$

(b) $\chi^2 = \dfrac{(13 - 38.6)^2}{38.6}$

$+ \dfrac{(18 - 22.1)^2}{22.1} + \dfrac{(25 - 12.4)^2}{12.4}$

$+ \dfrac{(38 - 10.3)^2}{10.3} + \dfrac{(6 - 16.5)^2}{16.5} = 111.72$

There are $5 - 1 = 4$ degrees of freedom. The critical value of χ^2 at the 0.05 level (from Appendix C) is 9.49. Since the calculated value of χ^2 is greater than the critical value, the null hypothesis can be rejected. There is a significant difference between the observed and expected frequencies at the 0.05 level.

4.7

The data can be set out in a 2 by 2 contingency table:

	Preference High	Preference Low	
Coastal	A 14	B 5	A+B 19
Inland	C 15	D 19	C+D 34
	A+C 29	B+D 24	n 53

$$\chi^2 = \frac{n\left(|AD - BC| - \dfrac{n}{2}\right)^2}{(A+B)(C+D)(A+C)(B+D)}$$

$$= \frac{53\left(|14 \times 19 - 5 \times 15| - \dfrac{53}{2}\right)^2}{19 \times 34 \times 29 \times 24} = 3.19$$

There is 1 degree of feedom. The critical value of χ^2 at the 0.05 level is 3.84. Since the calculated value of χ^2 is less than the critical value the null hypothesis can not be rejected. There is no significant difference in residential desirability between coastal and inland counties.

4.8, 4.9, 4.10, 5.1, *See pages 183–184*

5.2

(a) 0.492

(b) 0.708

(c) 15 individuals gives 13 degrees of freedom $(n-2)$. The two critical values are 0.441 (one-tailed) and 0.514 (two-tailed). The answers are therefore: (i) yes, (ii) no.

(d) The question simply asks whether there is or is not a significant correlation, without specifying a direction of correlation. A two-tailed test is appropriate. With $n - 2 = 8$ degrees of freedom, the critical value of r is 0.632 (from Appendix C8). Since the calculated value of r (0.702) is greater than the critical value in absolute terms, the null hypothesis can be rejected. There is a significant correlation between rainfall and altitude.

4.8

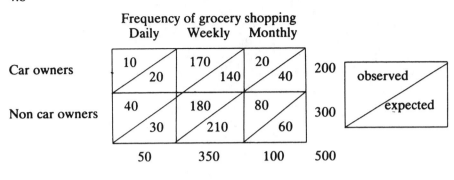

Frequency of grocery shopping

	Daily	Weekly	Monthly	
Car owners	10 / 20	170 / 140	20 / 40	200
Non car owners	40 / 30	180 / 210	80 / 60	300
	50	350	100	500

$$\chi^2 = \sum \frac{d^2}{e} = \frac{(10-20)^2}{20} + \frac{(170-140)^2}{140} + \frac{(20-40)^2}{40} + \frac{(40-30)^2}{30} + \frac{(180-210)^2}{210} + \frac{(80-60)^2}{60} = 35.71$$

There are 2 degrees of freedom. The critical value of χ^2 at the 0.01 level is 9.21. Since the calculated value of χ^2 is greater than the critical value the null hypothesis can be rejected. There is a significant difference in grocery shopping frequency between car owners and non car owners.

4.9 and 4.10

	A			B			C		
x	$(x-\bar{x})^2$	r_x	x	$(x-\bar{x})^2$	r_x	x	$(x-\bar{x})^2$	r_x
72	17.64	19	36	9.00	4.5	54	100.00	10
58	96.04	12	36	9.00	4.5	57	49.00	11
42	665.64	6	30	81.00	2	93	841.00	24
85	295.84	23	48	81.00	9	64	0.00	15
81	174.24	21	43	16.00	7	82	324.00	22
64	14.44	15	15	576.00	1	34	900.00	3
71	10.24	18	44	25.00	8			
68	0.04	17	60	441.00	13			
64	14.44	15						
73	27.04	20						
678	1315.60	166	312	1238.00	49	384	2214.00	85
	$n=10, \bar{x}=67.8$			$n=8, \bar{x}=39.0$			$n=6, \bar{x}=64.0$	
	$R=166$			$R=49$			$R=85$	

$$N = 10 + 8 + 6 = 24$$

$$H = \frac{12}{N(N+1)} \sum \frac{R^2}{n} - 3(N+1) = \frac{12}{24(24+1)} \left[\frac{166^2}{10} + \frac{49^2}{8} + \frac{85^2}{6} \right] - 3(24+1) = 10.198$$

There are 2 degrees of freedom. The critical value required is therefore the critical value of χ^2 at the 0.05 significance level, which is 5.99. Since the calculated value of H is greater than the critical value the null hypothesis can be rejected. There is a significant difference in water holding capacity between the three areas.

4.10

Grand mean $\bar{x}_G = \dfrac{678 + 312 + 384}{24} = 57.25$

$\hat{\sigma}_W^2 = \dfrac{\sum\limits^{k}\sum\limits^{n}(x - \bar{x})^2}{N - k}$

$= \dfrac{1315.60 + 1238.00 + 2214.00}{24 - 3} = 227.029$

$\hat{\sigma}_B^2 = \dfrac{\sum\limits^{k} n(\bar{x} - \bar{x}_G)^2}{k - 1}$

$= \dfrac{10(67.8 - 57.25)^2 + 8(39.0 - 57.25)^2 + 6(64.0 - 57.25)^2}{3 - 1}$

$= 2025.450$

$F = \dfrac{\hat{\sigma}_B^2}{\hat{\sigma}_W^2} = \dfrac{2025.450}{227.029} = 8.922$

The degrees of freedom are $2\,(k-1)$ for the larger variance estimate and $21\,(N-k)$ for the smaller variance estimate. The critical value of F at the 0.05 level with these degrees of freedom is 3.47. Since the calculated value of F is greater than the critical value, the null hypothesis can be rejected. There is a significant difference in water holding capacity between the three areas.

5.1

x (altitude)	y (rainfall)	x^2	y^2	xy
6.1	9.50	37.21	90.25	57.95
15.5	7.71	240.25	59.45	119.51
23.5	12.29	552.25	151.05	288.82
21.0	9.21	441.00	84.83	193.41
3.7	6.30	13.69	36.69	23.31
0.9	7.20	0.81	51.84	6.48
0.2	7.86	0.04	61.78	1.57
16.7	13.07	278.89	170.83	218.27
0.2	7.71	0.04	59.45	1.54
1.0	8.09	1.00	65.45	8.09
88.8	88.94	1565.18	834.59	918.95

5.1 continued

$\bar{x} = \dfrac{88.8}{10} = 8.88 \quad \bar{y} = \dfrac{88.94}{10} = 8.894$

$s_x = \sqrt{\dfrac{1565.18}{10} - 8.88^2} = 8.814$

$s_y = \sqrt{\dfrac{834.59}{10} - 8.894^2} = 2.087$

$r = \dfrac{\dfrac{918.95}{10} - 8.88 \times 8.894}{8.814 \times 2.087} = 0.702$

5.2 *See page 182*

5.3

Michael	John	d	d^2
1	6	−5	25
4	10	−6	36
8	2.5	5.5	30.25
7	4	3	9
9	1	8	64
3	7	−4	16
5	8	−3	9
2	5	−3	9
6	9	−3	9
10	2.5	7.5	56.25
n = 10			263.50

$r_s = 1 - \dfrac{6 \sum d^2}{n^3 - n}$

$= 1 - \dfrac{6 \times 263.50}{1000 - 10} = -0.597$

5.4

x	y	r_x	r_y	d	d^2
6.1	9.50	5	3	2	4
15.5	7.71	4	7.5	−3.5	12.25
23.5	12.29	1	2	−1	1
21.0	9.21	2	4	−2	4
3.7	6.30	6	10	−4	16
0.9	7.20	8	9	−1	1
0.2	7.86	9.5	6	3.5	12.25
16.7	13.07	3	1	2	4
0.2	7.71	9.5	7.5	2	4
1.0	8.09	7	5	2	4

n = 10					62.5

$$r_s = 1 - \frac{6 \sum d^2}{n^3 - n} = 1 - \frac{6 \times 62.5}{10^3 - 10} = 0.621$$

6.1

All the necessary totals for the calculation of the regression equation have already been calculated for Exercise 5.1.

$$b = \frac{\sum xy - n\bar{x}\bar{y}}{\sum x^2 - n\bar{x}^2}$$

$$= \frac{918.943 - (10 \times 8.88 \times 8.894)}{1565.180 - (10 \times 8.88^2)} = 0.166$$

$$a = \bar{y} - b\bar{x} = 8.88 - (0.166 \times 8.894) = 7.404$$

$$\therefore \hat{y} = 7.404 + 0.166x$$

6.2

$\hat{y}$	$\hat{y}^2$
8.42	70.84
9.98	99.54
11.31	127.80
10.89	118.59
8.02	64.29
7.55	57.05
7.44	55.31
10.18	103.56
7.44	55.31
7.57	57.31
88.78	809.60

$$s_y^2 = \frac{\sum y^2}{n} - \bar{y}^2 = \frac{834.589}{10} - 8.894^2 = 4.356$$

$$s_{\hat{y}}^2 = \frac{\sum \hat{y}^2}{n} - \bar{y}^2 = \frac{809.60}{10} - 8.894^2 = 1.858$$

$$r^2 = \frac{s_{\hat{y}}^2}{s_y^2} = \frac{1.858}{4.356} = 0.427$$

In other words, only about 42% of the variance of the dependent variable (rainfall) is accounted for by the regression.

6.3

y	$\hat{y}$	e	e^2
9.50	8.42	1.08	1.1664
7.71	9.98	−2.27	5.1529
12.29	11.31	0.99	0.9801
9.21	10.89	−1.68	2.8224
6.30	8.02	−1.72	2.9584
7.20	7.55	−0.35	0.1225
7.86	7.44	0.42	0.1764
13.07	10.18	2.89	8.3521
7.71	7.44	0.27	0.0729
8.09	7.57	0.52	0.2704
			22.0745

Sequence of residuals: + − + − − − + + + +

The number of runs in the sequence of residuals is 5. With $n_1 = 6$ and $n_2 = 4$, the critical values for the runs test at the 0.05 significance level are 2 and 9. Since the observed number of runs is not less than the lower critical value or greater than the higher critical value, the null hypothesis can not be rejected. The sequence of residuals can be considered to be random at the 0.05 level.

6.4

Residuals

a First nine	b Last nine	a^2	b^2	ab
1.08	2.27	1.1664	5.1529	2.4516
2.27	0.99	5.1529	0.9801	2.2473
0.99	1.68	0.9801	2.8224	1.6632
1.68	1.72	2.8224	2.9584	2.8896
1.72	0.35	2.9584	0.1225	0.6020
0.35	0.42	0.1225	0.1764	0.1470
0.42	2.89	0.1764	8.3521	1.2138
2.89	0.27	8.3521	0.0729	0.7803
0.27	0.52	0.0729	0.2704	0.1404
11.67	11.11	21.8041	20.9081	12.1352

$$\bar{a} = \frac{11.67}{9} = 1.297 \quad \bar{b} = \frac{11.11}{9} = 1.234$$

$$s_a = \sqrt{\frac{21.8041}{9} - 1.297^2} = 0.861$$

$$s_b = \sqrt{\frac{20.9081}{9} - 1.234^2} = 0.895$$

$$r = \frac{\dfrac{12.1352}{9} - (1.297 \times 1.234)}{0.861 \times 0.895} = -0.327$$

6.5

(a) $x_0 = 65$, confidence interval

$$= 2.31 \sqrt{33.07 \left[0.1 + \frac{(65-50)^2}{2818}\right]} = 5.63$$

(b) 3.35

(c) From Equation 6.5:

when

$$x_0 = 50, \hat{y} = -2.6 + (0.652 \times 50) = 30.00$$

when

$$x_0 = 60, \hat{y} = -2.6 + (0.652 \times 60) = 36.52$$

when

$$x_0 = 65, \hat{y} = -2.6 + (0.652 \times 65) = 39.78$$

(i) Confidence interval

$$= 3.35 \sqrt{33.07 \left[0.1 + \frac{(50-50)^2}{2818}\right]}$$
$$= 6.09$$

99% limits are set by $30.00 \pm 6.09 =$ 23.91 and 36.09

(ii) Confidence interval

$$= 3.35 \sqrt{33.07 \left[0.1 + \frac{(60-50)^2}{2818}\right]}$$
$$= 7.09$$

99% limits are set by $36.52 \pm 7.09 =$ 29.43 and 43.61

(iii) Confidence interval

$$3.35 \sqrt{33.07 \left[0.1 + \frac{(65-50)^2}{2818}\right]} = 8.17$$

99% limits are set by $39.78 \pm 8.17 =$ 31.61 and 47.95

6.6
The appropriate value of t for the 95% confidence interval is 2.31 (from Appendix C4). The equation for the confidence interval is:

$$\text{confidence interval} = t \sqrt{\frac{\sum e^2}{n-2} \left[\frac{1}{n} + \frac{(x_0 - \bar{x})^2}{\sum x^2 - n\bar{x}^2}\right]}$$

The necessary elements of this equation have already been calculated for previous exercises, so in this case:

confidence interval

$$= 2.31 \sqrt{\frac{22.0745}{10-2} \left[\frac{1}{10} + \frac{(x_0 - 8.88)^2}{1565.18 - (10 \times 8.88^2)}\right]}$$

$x_0 = 6.1$ confidence interval $= 1.272$
$x_0 = 15.5$ confidence interval $= 1.518$
$x_0 = 23.5$ confidence interval $= 2.350$
$x_0 = 21.0$ confidence interval $= 2.063$
$x_0 = 3.7$ confidence interval $= 1.408$
$x_0 = 0.9$ confidence interval $= 1.637$
$x_0 = 0.2$ confidence interval $= 1.703$
$x_0 = 16.7$ confidence interval $= 1.622$
$x_0 = 0.2$ confidence interval $= 1.703$
$x_0 = 1.0$ confidence interval $= 1.628$

6.7

(a) $\hat{y} = \dfrac{11750}{50^{2.06}} = 3.716$

(b) $\hat{y} = \dfrac{11750}{75^{2.06}} = 1.612$

6.8

(a) Sequence of residuals: $- - - + + + - +$
$- + + + + + - - - - + +$
Number of runs $= 8$, $n_1 = 9$, $n_2 = 11$. Critical values at the 0.05 level are 6 and 16 (from Appendix C2). Since the observed number of runs is not less than the lower critical value or greater than the higher critical value, the null hypothesis of randomness can not be rejected. The sequence of residuals can be considered to be random at the 0.05 significance level.

(b)

Residuals

a	b	a^2	b^2	ab
First 19	Last 19			
0.0027	0.1235	0.0000	0.0153	0.0003
0.1235	0.1421	0.0153	0.0202	0.0175
0.1421	0.1664	0.0202	0.0277	0.0236
0.1664	0.0401	0.0277	0.0016	0.0067
0.0401	0.0867	0.0016	0.0075	0.0035
0.0867	0.0441	0.0075	0.0020	0.0038
0.0441	0.0085	0.0020	0.0001	0.0004
0.0085	0.0623	0.0001	0.0039	0.0005
0.0623	0.0320	0.0039	0.0010	0.0020
0.0320	0.2142	0.0010	0.0459	0.0069
0.2142	0.0700	0.0459	0.0049	0.0150
0.0700	0.1417	0.0049	0.0201	0.0099
0.1417	0.0319	0.0201	0.0010	0.0045
0.0319	0.2073	0.0010	0.0430	0.0066
0.2073	0.1496	0.0430	0.0224	0.0310
0.1496	0.0954	0.0224	0.0091	0.0143
0.0954	0.0443	0.0091	0.0020	0.0042
0.0443	0.0040	0.0020	0.0000	0.0002
0.0040	0.0499	0.0000	0.0025	0.0002
1.6668	1.7140	0.2276	0.2301	0.1511

$$\bar{a} = \frac{1.6668}{19} = 0.0877 \quad \bar{b} = \frac{1.7140}{19} = 0.0902$$

$$s_a = \sqrt{\frac{0.2276}{19} - 0.0877^2} = 0.0655$$

$$s_b = \sqrt{\frac{0.2301}{19} - 0.0902^2} = 0.0630$$

$$r = \frac{\frac{0.1511}{19} - (0.0877 \times 0.0902)}{0.0655 \times 0.0630} = 0.01$$

(c) $s_{\hat{y}}^2 = \frac{\sum \hat{y}'^2}{n} - \bar{y}'^2 = \frac{20.8886}{20} - 0.735^2$
$= 0.5042$

$s_{y'}^2 = \frac{\sum y'^2}{n} - \bar{y}'^2 = \frac{20.9834}{20} - 0.735^2 = 0.5090$

$r^2 = \frac{s_{\hat{y}}^2}{s_{y'}^2} = \frac{0.5042}{0.5090} = 0.991$

The regression accounts for 99.1% of the variance in the dependent variable.

7.1   See page 188

7.2
The true location of the centre of minimum travel (found by computer) has the co-ordinates 8.56, 16.50. This puts it about 3 miles northwest of Hucknall.

7.3
(a)

Point	Co-ordinates			
	x	y	x^2	y^2
A	2	5	4	25
B	1	4	1	16
C	3	2	9	4
D	4	1	16	1
E	5	1	25	1
n = 5	15	13	55	47

$$\bar{x} = \frac{15}{5} = 3.0, \bar{y} = \frac{13}{5} = 2.6$$

standard distance

$$= \sqrt{\left(\frac{\sum x^2}{n} - \bar{x}^2\right) + \left(\frac{\sum y^2}{n} - \bar{y}^2\right)}$$

$$= \sqrt{\left(\frac{55}{5} - 3.0^2\right) + \left(\frac{47}{5} - 2.6^2\right)} = 2.154$$

(b) Standard distance = 10.04

7.4, 7.5   See page 189

7.6

Nearest neighbour distances

A	2.236
B	1.000
C	1.000
D	1.414
E	1.414
F	1.414
G	2.236
H	1.414
I	2.000
J	1.414
n = 10	15.542

7.1

Town	Co-ordinates		Weight	Weighted co-ordinates	
	x	y	w	xw	yw
East Retford	17	34	18	306	612
Worksop	9	33	21	189	693
Warsop	8	27	36	288	972
Ollerton	14	26	30	420	780
Mansfield Woodhouse	7	23	25	175	575
Mansfield	6	22	58	348	1276
Sutton in Ashfield	4	21	41	164	861
Kirkby in Ashfield	3	19	24	72	456
Southwell	16	17	27	432	459
Newark	23	17	25	575	425
Hucknall	6	15	26	156	390
Eastwood	2	13	11	22	143
Arnold	10	12	33	330	396
Carlton	10	10	45	450	450
Nottingham	7	9	300	2100	2700
Bingham	17	8	39	663	312
Beeston	4	7	63	252	441
West Bridgford	9	6	28	252	168
n = 18	172	319	850	7194	12109

(a) $\bar{x} = \dfrac{172}{18} = 9.56$, $\bar{y} = \dfrac{319}{18} = 17.72$

Co-ordinates of mean centre are therefore: 9.56, 17.72. This puts the mean centre about 6 miles due north of Arnold.

(b) $\bar{x}_w = \dfrac{7194}{850} = 8.46$, $\bar{y}_w = \dfrac{12109}{850} = 14.25$

Co-ordinates of weighted mean centre are therefore: 8.46, 14.25. The weighted mean centre is about 2 miles northwest of Arnold.

7.6 continued

Mean nearest neighbour distance

$$\bar{d}_{obs} = \frac{15.542}{10} = 1.554$$

Total area of rectangle enclosing all the points = $8 \times 8 = 64$

$$\therefore \text{ point density } p = \frac{10}{64} = 0.156$$

$$\bar{d}_{ran} = \frac{1}{2\sqrt{p}} = \frac{1}{2\sqrt{0.156}} = 1.266$$

$$R = \frac{\bar{d}_{obs}}{\bar{d}_{ran}} = \frac{1.554}{1.266} = 1.227$$

With 10 points the critical value of R at the 0.01

level for a one-tailed test (from Appendix C11a) is 0.615. Since the calculated value of R is not less than the critical value the null hypothesis can not be rejected. The pattern is not significantly clustered at the 0.01 level. A one-tailed test has been chosen because a direction of departure from randomness has been specified.

7.7

(a) The observed mean nearest-neighbour distance ($\bar{d}_{obs}$) is 0.072 km. With 13 points in an area of 0.29 sq. km. the point density is $13/0.29 = 44.8$. The expected nearest neighbour distance for a random pattern ($\bar{d}_{ran}$) is $1/2 \sqrt{44.8} = 0.075$. The value of the nearest-neighbour index, R, is therefore $0.072/0.075 =$

Continued on page 190

7.4

	Co-ordinates			Transposed co-ordinates			
Point	x	y	x'	y'	x'^2	y'^2	x'y'
A	1.2	4.0	−1.475	1.450	2.1756	2.1025	−2.1388
B	1.8	3.2	−0.875	0.650	0.7656	0.4225	−0.5688
C	2.7	3.0	0.025	0.450	0.0006	0.2025	0.0113
D	3.7	3.2	1.025	0.650	1.0506	0.4225	0.6663
E	2.3	2.2	−0.375	−0.350	0.1406	0.1225	0.1313
F	3.6	2.1	0.925	−0.450	0.8556	0.2025	0.4163
G	3.2	1.7	0.525	−0.850	0.2756	0.7225	−0.4463
H	2.9	1.0	0.225	−1.550	0.0506	2.4025	−0.3488
n = 8					5.3148	6.6000	−2.2775

$$\tan \theta = \frac{(5.3148 - 6.6000) + \sqrt{(5.3148 - 6.6000)^2 + 4(-2.2775)^2}}{2(-2.2775)} = -0.7565$$

∴ $\theta = 90 - 37° 8'$ (since $\tan \theta$ is negative, the angle must be subtracted from 90° to give the correct value of θ) = 52° 52'
$\sin \theta = 0.797$, $\cos \theta = 0.604$, $\sin^2\theta = 0.635$, $\cos^2\theta = 0.365$, $\sin \theta \cos \theta = 0.481$

$$\therefore \sigma_x = \sqrt{\frac{(5.3148)0.365 - 2(-2.2775)0.481 + (6.6000)0.635}{8}} = 1.020$$

$$\sigma_y = \sqrt{\frac{(5.3148)0.635 + 2(-2.2775)0.481 + (6.6000)0.365}{8}} = 0.670$$

7.5
(a)

	A	R_A	R_C	D	R_I	L	S_2	S_3	S_4	S_5
Yorkshire East Riding	1015	18.0	28	56	15	55	0.427	0.412	0.643	0.536
Humberside	1345	20.7	28	56	12	57	0.527	0.546	0.739	0.429
Warwickshire (before)	980	17.7	25	50	13	51	0.480	0.499	0.708	0.520
Warwickshire (after)	795	12.0	25	50	10	51	0.389	0.405	0.480	0.400
Gloucestershire (before)	1250	19.9	30	60	12	60	0.442	0.442	0.663	0.400
Gloucestershire (after)	840	16.4	26	52	12	62	0.278	0.396	0.631	0.462

A = area in square miles
R_A = radius of equal area circle in miles
R_C = radius of smallest circumscribing circle in miles
D = diameter of smallest circumscribing circle in miles

R_I = radius of largest inscribed circle in miles
L = length of longest axis in miles in miles

(b) Yes and no! It depends which measure of shape is employed.

7.7 *continued*

0.960. A one-tailed test is appropriate, since a direction of departure from randomness is specified. The one-tailed critical value at the 0.05 level for a clustered pattern (from Appendix C11a) is 0.762. As the calculated value of R is greater than the critical value, the null hypothesis can be accepted. The pattern is not significantly clustered at the 0.05 level.

(b) Not really. Some of the assumptions of the technique are violated. Houses for sale are not free to locate anywhere within the area, since the positions of all the houses are fixed. One would also want to know how the study area had been defined, and whether it is relatively homogeneous in terms of the type of housing. Are all the houses equally likely to be put up for sale? Are they all in the same general price range?

7.8

(a) For Figure 7.21b $O_{BW} = 12$, $E_{BW} = 9.12$ and $\sigma_{BM} = 2.416$. The last two values have already been calculated for Figure 7.21a, which has the same spatial structure as Figures 7.21b and c.

$$\therefore z = \frac{O_{BW} - E_{BW}}{\sigma_{BW}} = \frac{12 - 9.12}{2.416} = 1.192$$

A two-tailed test is appropriate here since the question only asks whether the pattern is significantly different from a random pattern, without specifying a direction of departure from randomness. The critical value of z at the 0.05 level for a two-tailed test (from Appendix C10) is 1.960. As the calculated value of z is less than the critical value, the null hypothesis of randomness can not be rejected. The pattern does not depart significantly from randomness at the 0.05 level. For Figure 7.21c $O_{BW} = 10$, $E_{BW} = 9.12$ and $\sigma_{BW} = 2.416$.

$$\therefore z = \frac{10 - 9.12}{2.416} = 0.364$$

The critical value of z is again 1.960. As the calculated z is less than the critical value, the null hypothesis can not be rejected at the 0.05 level.

(b) $O_{BW} = 6$, $E_{BW} = 2Jpq$

$$= 2 \times 19 \times 0.45 \times 0.55 = 9.405$$

$$\sigma_{BW} = \sqrt{\begin{array}{c}(38 + 114)0.45 \times 0.55 \\ -4(19 + 114)0.45^2 \times 0.55^2\end{array}}$$

$$= \sqrt{37.620 - 32.588} = 2.243$$

$$z = \frac{O_{BW} - E_{BW}}{\sigma_{BW}} = \frac{6 - 9.405}{2.243} = -0.518$$

A one-tailed test is appropriate since a direction of autocorrelation is specified. The critical value of z for a one-tailed test at the 0.05 level is -1.645. Since the calculated value of z is not less than the critical value (remember that this is a negative value of z), the null hypothesis can not be rejected. The pattern is not significantly clustered at the 0.05 level.

7.9

$O_{BW} = 12$, $E_{BW} = 10.3636$ and $\sigma_{BW} = 1.7721$.

$$\therefore z = \frac{12 - 10.3636}{1.7721} = 0.923$$

Since no direction of autocorrelation is specified, a two-tailed test is appropriate. The critical value of z at the 0.01 level is 2.576. The calculated value of z is less than the critical value, so the null hypothesis can not be rejected. The pattern is not significantly different from random at the 0.01 level.

7.10

Either form of null hypothesis is possible. For the free sampling null hypothesis it is possible to use the national figures to calculate the probability of a 'black' area (State with a Republican majority of the popular vote). If the total number of votes cast for the two main parties' candidates is 79.979m (39.148m + 40.831m), p = 39.148/79.979 = 0.489478 and

$q = 1 - 0.489478 = 0.510522$. However, I would feel happier with the non-free sampling null hypothesis. Calculations for both versions are given below.

$O_{BW} = 32$, $B = 26$, $W = 22$, $n = 48$, $J = 107$, $p = 0.489478$, $q = 0.510522$

	L	L–1	L(L–1)		L	L–1	L(L–1)
Alabama	4	3	12	Nebraska	6	5	30
Arizona	5	4	20	Nevada	5	4	20
Arkansas	6	5	30	New Hampshire	3	2	6
California	3	2	6	New Jersey	3	2	6
Colorado	7	6	42	New Mexico	5	4	20
Connecticut	3	2	6	New York	5	4	20
Delaware	3	2	6	North Carolina	4	3	12
Florida	2	1	2	North Dakota	3	2	6
Georgia	5	4	20	Ohio	5	4	20
Idaho	6	5	30	Oklahoma	6	5	30
Illinois	5	4	20	Oregon	4	3	12
Indiana	4	3	12	Pennsylvania	6	5	30
Iowa	6	5	30	Rhode Island	2	1	2
Kansas	4	3	12	South Carolina	2	1	2
Kentucky	7	6	42	South Dakota	6	5	30
Louisiana	3	2	6	Tennessee	8	7	56
Maine	1	0	0	Texas	4	3	12
Maryland	4	3	12	Utah	6	5	30
Massachusetts	5	4	20	Vermont	3	2	6
Michigan	3	2	6	Virginia	5	4	20
Minnesota	4	3	12	Washington	2	1	2
Mississippi	4	3	12	West Virginia	5	4	20
Missouri	8	7	56	Wisconsin	4	3	12
Montana	4	3	12	Wyoming	6	5	30

$\Sigma L(L{-}1) = 860$

Free sampling null hypothesis:

$E_{BW} = 2Jpq = 2 \times 107 \times 0.489478 \times 0.510522 = 53.4763$

$$\sigma_{BW} = \sqrt{[2J + \Sigma L(L - 1)]pq - 4[J + \Sigma L(L - 1)]p^2q^2}$$

$$= \sqrt{[(2 \times 107) + 860]0.489478 \times 0.510522 - 4(107 + 860)0.489478^2 \times 0.510522^2}$$

$$= \sqrt{268.3811 - 241.5359} = \sqrt{26.8452} = 5.1812$$

$$z \text{ (standard normal deviate)} = \frac{O_{BW} - E_{BW}}{\sigma_{BW}} = \frac{32 - 53.4763}{5.1812} = -4.1450$$

Non-free sampling null hypothesis:

$$E_{BW} = \frac{2JBW}{n(n-1)} = \frac{2 \times 107 \times 26 \times 22}{48 \times 47} = 54.2589$$

$$\sigma_{BW} = \sqrt{E_{BW} + \frac{\Sigma L(L-1)BW}{n(n-1)} + \frac{4[J(J-1) - \Sigma L(L-1)]B(B-1)W(W-1)}{n(n-1)(n-2)(n-3)} - E^2_{BW}}$$

$$= \sqrt{54.2589 + \frac{860 \times 26 \times 22}{48 \times 47} + \frac{4[(107 \times 106) - 860] \times 26 \times 25 \times 22 \times 21}{48 \times 47 \times 46 \times 45} - 54.2589^2}$$

$$= \sqrt{54.2589 + 218.0496 + 269.1872 - 2944.0245} = \sqrt{24.4712} = 4.9468$$

$$z \text{ (standard normal deviate)} = \frac{O_{BW} - E_{BW}}{\sigma_{BW}} = \frac{32 - 54.2589}{4.9468} = -4.4996$$

The critical value of a standard normal deviate for a one-tailed test at the 0.01 probability level is -2.326 (from Appendix C10). The calculated standard normal deviate in both cases is of the correct sign and larger in absolute value than the critical value. Both null hypotheses can be rejected at the 0.01 level. There is a significant clustering of 'Republican States'.

7.11

The randomization null hypothesis is appropriate in this case. It would not be realistic to consider the percentage Labour vote in the 18 constituencies as a sample of values from a normal distribution. All that it is possible to do is to consider whether it is probable that the observed spatial arrangement of votes occurred by chance.

	L	L^2	x	$(x-\bar{x})$	$(x-\bar{x})^2$	$(x-\bar{x})^4$
Battersea N	3	9	38.3	3.3	10.89	118.59
Battersea S	4	16	30.7	−4.3	18.49	341.88
Bermondsey	3	9	41.4	6.4	40.96	1677.72
Deptford	6	36	34.6	−0.4	0.16	0.03
Dulwich	5	25	32.3	−2.7	7.29	53.14
Erith & Crayford	1	1	37.4	2.4	5.76	33.18
Greenwich	4	16	36.3	1.3	1.69	2.86
Lambeth	6	36	31.6	−3.4	11.56	133.63
Lewisham E	4	16	35.0	0.0	0.00	0.00
Lewisham W	3	9	33.8	−1.2	1.44	2.07
Mitcham & Morden	1	1	34.2	−0.8	0.64	0.41
Norwood	3	9	31.1	−3.9	15.21	231.34
Peckham	6	36	39.1	4.1	16.81	282.58
Putney	1	1	32.5	−2.5	6.25	39.06
Tooting	2	4	34.5	−0.5	0.25	0.06
Vauxhall	4	16	33.3	−1.7	2.89	8.35
Woolwich E	3	9	38.9	3.9	15.21	231.34
Woolwich W	3	9	34.8	−0.2	0.04	0.00
	258	629.8			155.54	3156.24

$$\bar{x} = \frac{\sum x}{n} = \frac{629.8}{18} = 35.0$$

$$\text{kurtosis (k)} = \frac{\sum (x-\bar{x})^4}{n\sigma^4}$$

$$= \frac{\sum (x-\bar{x})^4}{n\left[\frac{\sum (x-\bar{x})^2}{n}\right]^2} = \frac{3156.24}{18\left[\frac{155.54}{18}\right]^2}$$

$$= \frac{3156.24}{18 \times 8.6411^2} = \frac{3156.24}{18 \times 74.67}$$

$$= \frac{3156.24}{1344.06} = 2.348$$

Area i	$(x_i - \bar{x})$	Area j	$(x_j - \bar{x})$	$(x_i - \bar{x})(x_j - \bar{x})$
Battersea N	3.3	Battersea S	−4.3	−14.19
		Lambeth	−3.4	−11.22
		Vauxhall	−1.7	−5.61
Battersea S	−4.3	Lambeth	−3.4	14.62
		Putney	−2.5	10.75
		Tooting	−0.5	2.15
Bermondsey	6.4	Deptford	−0.4	−2.56
		Peckham	4.1	26.24
		Vauxhall	−1.7	−10.88
Deptford	−0.4	Dulwich	−2.7	1.08
		Greenwich	1.3	−0.52
		Lewisham E	0.0	0.00
		Lewisham W	−1.2	0.48
		Peckham	4.1	−1.64
Dulwich	−2.7	Lambeth	−3.4	9.18
		Lewisham W	−1.2	3.24
		Norwood	−3.9	10.53
		Peckham	4.1	−11.07
Erith & Crayford	2.4	Woolwich E	3.9	9.36
Greenwich	1.3	Lewisham E	0.0	0.00
		Woolwich E	3.9	5.07
		Woolwich W	−0.2	−0.26
Lambeth	−3.4	Norwood	−3.9	13.26
		Peckham	4.1	−13.94
		Vauxhall	−1.7	5.78
Mitcham & Morden	−0.8	Tooting	−0.5	0.40
Norwood	−3.9	Peckham	4.1	−15.99
Peckham	4.1	Vauxhall	−1.7	−6.97
Lewisham E	0.0	Lewisham W	−1.2	0.00
		Woolwich W	−0.2	0.00
Woolwich E	3.9	Woolwich W	−0.2	−0.78

$$\sum_{(c)} (x_i - \bar{x})(x_j - \bar{x}) = 16.51$$

$$I = \frac{n \sum_{(c)} (x_i - \bar{x})(x_j - \bar{x})}{J \sum (x - \bar{x})^2} = \frac{18 \times 16.51}{31 \times 155.54} = \frac{297.18}{4821.74} = 0.0616$$

$$E_I = -\frac{1}{n-1} = -\frac{1}{17} = -0.0588$$

$$\sigma_I = \sqrt{\frac{n[J(n^2 + 3 - 3n) + 3J^2 - n \sum L^2]}{-k[J(n^2 - n) + 6J^2 - 2n \sum L^2]}{J^2(n-1)(n-2)(n-3)}}$$

$$= \sqrt{\frac{18[31(18^2 + 3 - 3 \times 18) + 3 \times 31^2 - 18 \times 258]}{-2.348[31(18^2 - 18) + 6 \times 31^2 - 2 \times 18 \times 258]}{31^2(18-1)(18-2)(18-3)}}$$

$$= \sqrt{\frac{18[8463 + 2883 - 4644] - 2.348[9486 + 5766 - 9288]}{961 \times 17 \times 16 \times 15}} = \sqrt{\frac{120636 - 14003.472}{3920880}}$$

$$= \sqrt{\frac{106632.528}{3920880}} = \sqrt{0.0272} = 0.165$$

$$z = \frac{I - E_I}{\sigma_I} = \frac{0.0616 - (-0.0588)}{0.165} = \frac{0.1204}{0.165} = 0.730$$

Since we are not testing for any specific **direction** of spatial autocorrelation in this case, a two-tailed test is appropriate. The two-tailed critical value of a standard normal deviate at the 0.05 level is 1.960 (from Appendix C10). The calculated standard normal deviate is smaller than the critical value so the null hypothesis can be accepted. There is no significant spatial autocorrelation at the 0.05 level.

Probability Tables

B1 Normal Probabilities

Entries in the body of the table give the area under a normal distribution curve **to the left** of specified units of standard deviation. The total area under the curve is 1000. The probability that an individual in a normal distribution has a value less than a specified number of standard deviates is given by dividing the appropriate entry in the table by 1000. For example, the probability that an individual in a normal distribution will have a value less than 1.23 standard deviates is $891/1000 = 0.891$.

Standard Deviates

	.0	.1	.2	.3	.4	.5	.6	.7	.8	.9
−4	0.032	0.021	0.013	0.009	0.0054	0.0034	0.0021	0.0013	0.0003	0.0000
−3	1.35	0.97	0.69	0.48	0.34	0.23	0.16	0.11	0.07	0.05
−2	22.8	17.9	13.9	10.7	8.2	6.2	4.7	3.5	2.56	1.87

	.00	.01	.02	.03	.04	.05	.06	.07	.08	.09
−1.9	29	28	27	27	26	26	25	24	24	23
−1.8	36	35	34	34	33	32	31	31	30	29
−1.7	45	44	43	42	41	40	39	38	38	37
−1.6	55	54	53	52	51	49	48	47	46	46
−1.5	67	66	64	63	62	61	59	58	57	56
−1.4	81	79	78	76	75	74	72	71	69	68
−1.3	97	95	93	92	90	89	87	85	84	82
−1.2	115	113	111	109	107	106	104	102	100	99
−1.1	136	133	131	129	127	125	123	121	119	117
−1.0	159	156	154	152	149	147	145	142	140	138
−0.9	184	181	179	176	174	171	169	166	164	161
−0.8	212	209	206	203	200	198	195	192	189	187
−0.7	242	239	236	233	230	227	224	221	218	215
−0.6	274	271	268	264	261	258	255	251	248	245
−0.5	309	305	302	298	291	295	288	284	281	278
−0.4	345	341	337	334	330	326	323	319	316	312
−0.3	382	378	374	371	367	363	359	356	352	348
−0.2	421	417	413	409	405	401	397	394	390	386

continued

	.00	.01	.02	.03	.04	.05	.06	.07	.08	.09
−0.1	460	456	452	448	444	440	436	433	429	425
−0.0	500	496	492	488	484	480	476	472	468	464
+0.0	500	504	508	512	516	520	524	528	532	536
+0.1	540	544	548	552	556	560	564	567	571	575
+0.2	579	583	587	591	595	599	603	606	610	614
+0.3	618	622	626	629	633	637	641	644	648	652
+0.4	655	659	663	666	670	674	677	681	684	688
+0.5	691	695	698	702	705	709	712	716	719	722
+0.6	726	729	732	736	739	742	745	749	752	755
+0.7	758	761	764	767	770	773	776	779	782	785
+0.8	788	791	794	797	800	802	805	808	811	813
+0.9	816	819	821	824	826	829	831	834	836	839
+1.0	841	844	846	848	851	853	855	858	860	862
+1.1	864	867	869	871	873	875	877	879	881	883
+1.2	885	887	889	891	893	894	896	898	900	901
+1.3	903	905	907	908	910	911	913	915	916	918
+1.4	919	921	922	924	925	926	928	929	931	932
+1.5	933	934	936	937	938	939	941	942	943	944
+1.6	945	946	947	948	949	951	952	953	954	954
+1.7	955	956	957	958	959	960	961	962	962	963
+1.8	964	965	966	966	967	968	969	969	970	971
+1.9	971	972	973	973	974	974	975	976	976	977

	.0	.1	.2	.3	.4	.5	.6	.7	.8	.9
+2	977.2	982.1	986.1	989.3	991.8	993.8	995.3	996.5	997.44	998.13
+3	998.65	999.03	999.31	999.52	999.66	999.77	999.84	999.89	999.93	999.95
+4	999.968	999.979	999.987	999.991	999.9946	999.9966	999.9979	999.9987	999.9997	1000.0

B2 Values of e^z for Use in Calculating Poisson Probabilities

z	e^z	z	e^z	z	e^z	z	e^z
0.01	1.0101	0.13	1.1388	0.25	1.2840	0.37	1.4477
0.02	1.0202	0.14	1.1503	0.26	1.2969	0.38	1.4623
0.03	1.0305	0.15	1.1618	0.27	1.3100	0.39	1.4770
0.04	1.0408	0.16	1.1735	0.28	1.3231	0.40	*1.4918
0.05	1.0513	0.17	1.1853	0.29	1.3364	0.41	1.5068
0.06	1.0618	0.18	1.1972	0.30	1.3499	0.42	1.5220
0.07	1.0725	0.19	1.2093	0.31	1.3634	0.43	1.5373
0.08	1.0833	0.20	1.2214	0.32	1.3771	0.44	1.5527
0.09	1.0942	0.21	1.2337	0.33	1.3910	0.45	1.5683
0.10	1.1052	0.22	1.2461	0.34	1.4050	0.46	1.5841
0.11	1.1163	0.23	1.2586	0.35	1.4191	0.47	1.6000
0.12	1.1275	0.24	1.2713	0.36	1.4333	0.48	1.6161

z	e^z	z	e^z	z	e^z	z	e^z
0.49	1.6323	0.87	2.3869	3.5	33.1155	7.3	1480.2991
0.50	1.6487	0.88	2.4109	3.6	36.5982	7.4	1635.9835
0.51	1.6653	0.89	2.4351	3.7	40.4473	7.5	1808.0415
0.52	1.6820	0.90	2.4596	3.8	44.7012	7.6	1998.1949
0.53	1.6989	0.91	2.4843	3.9	49.4024	7.7	2208.3464
0.54	1.7160	0.92	2.5093	4.0	54.5981	7.8	2440.6003
0.55	1.7333	0.93	2.5345	4.1	60.3403	7.9	2697.2805
0.56	1.7507	0.94	2.5600	4.2	66.6863	8.0	2980.9561
0.57	1.7683	0.95	2.5857	4.3	73.6998	8.1	3294.4660
0.58	1.7860	0.96	2.6117	4.4	81.4509	8.2	3640.9481
0.59	1.8040	0.97	2.6380	4.5	90.0171	8.3	4023.8701
0.60	1.8221	0.98	2.6645	4.6	99.4843	8.4	4447.0633
0.61	1.8404	0.99	2.6912	4.7	109.9472	8.5	4914.7651
0.62	1.8589	1.00	2.7183	4.8	121.5104	8.6	5431.6556
0.63	1.8776	1.1	3.0042	4.9	134.2897	8.7	6002.9080
0.64	1.8965	1.2	3.3201	5.0	148.4131	8.8	6634.2394
0.65	1.9156	1.3	3.6693	5.1	164.0218	8.9	7331.9686
0.66	1.9348	1.4	4.0552	5.2	181.2722	9.0	8103.0787
0.67	1.9542	1.5	4.4817	5.3	200.3367	9.1	8955.2851
0.68	1.9739	1.6	4.9530	5.4	221.4063	9.2	9897.1209
0.69	1.9937	1.7	5.4740	5.5	244.6919	9.3	10938.0105
0.70	2.0138	1.8	6.0497	5.6	270.4263	9.4	12088.3713
0.71	2.0340	1.9	6.6859	5.7	298.8673	9.5	13359.7167
0.72	2.0544	2.0	7.3891	5.8	330.2994	9.6	14764.7707
0.73	2.0751	2.1	8.1662	5.9	365.0373	9.7	16317.5955
0.74	2.0959	2.2	9.0250	6.0	403.4286	9.8	18033.7285
0.75	2.1170	2.3	9.9742	6.1	445.8576	9.9	19930.3527
0.76	2.1383	2.4	11.0232	6.2	492.7489	10.0	22026.4467
0.77	2.1598	2.5	12.1825	6.3	544.5716	10.1	24342.9888
0.78	2.1815	2.6	13.4637	6.4	601.8447	10.2	26903.1639
0.79	2.2034	2.7	14.8797	6.5	665.1413	10.3	29732.5949
0.80	2.2256	2.8	16.4447	6.6	735.0948	10.4	32859.5927
0.81	2.2479	2.9	18.1742	6.7	812.4054	10.5	36315.4671
0.82	2.2705	3.0	20.0855	6.8	897.8469	10.6	40134.7990
0.83	2.2933	3.1	22.1980	6.9	992.2743	10.7	44355.8136
0.84	2.3164	3.2	24.5325	7.0	1096.6325	10.8	49020.7564
0.85	2.3397	3.3	27.1126	7.1	1211.9663	10.9	54176.3154
0.86	2.3632	3.4	29.9641	7.2	1339.4300	11.0	59874.0895

Tables of Critical Values

C1 Critical Values of D for the Kolmogorov–Smirnov Goodness of Fit test

Degrees of freedom	Significance level				
	0.20	0.15	0.10	0.05	0.01
1	0.900	0.925	0.950	0.975	0.995
2	0.684	0.726	0.776	0.842	0.929
3	0.565	0.597	0.642	0.708	0.829
4	0.494	0.525	0.564	0.624	0.734
5	0.446	0.474	0.510	0.563	0.669
6	0.410	0.436	0.470	0.521	0.618
7	0.381	0.405	0.438	0.486	0.577
8	0.358	0.381	0.411	0.457	0.543
9	0.339	0.360	0.388	0.432	0.514
10	0.322	0.342	0.368	0.409	0.486
11	0.307	0.326	0.352	0.391	0.468
12	0.295	0.313	0.338	0.375	0.450
13	0.284	0.302	0.325	0.361	0.433
14	0.274	0.292	0.314	0.349	0.418
15	0.266	0.283	0.304	0.338	0.404
16	0.258	0.274	0.295	0.328	0.391
17	0.250	0.266	0.286	0.318	0.380
18	0.244	0.259	0.278	0.309	0.370
19	0.237	0.252	0.272	0.301	0.361
20	0.231	0.246	0.264	0.294	0.352
25	0.21	0.22	0.24	0.264	0.32
30	0.19	0.20	0.22	0.242	0.29
35	0.18	0.19	0.21	0.23	0.27
40	0.17	0.18	0.19	0.21	0.25
50	0.15	0.16	0.17	0.19	0.23
60	0.14	0.15	0.16	0.17	0.21
70	0.13	0.14	0.15	0.16	0.19
80	0.12	0.13	0.14	0.15	0.18
90	0.11	0.12	0.13	0.14	0.17
100	0.11	0.11	0.12	0.14	0.16

For degrees of freedom greater than 100, $D_{crit} =$

$$\frac{1.07}{\sqrt{n}} \qquad \frac{1.14}{\sqrt{n}} \qquad \frac{1.22}{\sqrt{n}} \qquad \frac{1.36}{\sqrt{n}} \qquad \frac{1.63}{\sqrt{n}}$$

Reject H_0 if calculated value of D is **greater than** the critical value at the chosen significance level

C2 Critical Values of the Number of Runs for the Runs Test at the 0.05 Significance Level

n_2 \ n_1	2	3	4	5	6	7	8	9	10	11
2										
3					2	2	2	2	2	2
4				2/9	2/9	2	3	3	3	3
5		2	2/9	2/10	3/10	3/11	3/11	3	3	4
6		2	2/9	3/10	3/11	3/12	3/12	4/13	4/13	4/13
7		2	2	3/11	3/12	3/13	4/13	4/14	5/14	5/14
8		2	3	3/11	3/12	4/13	4/14	5/14	5/15	5/15
9		2	3	3	4/13	4/14	5/14	5/15	5/16	6/16
10		2	3	3	4/13	5/14	5/15	5/16	6/16	6/17
11		2	3	4	4/13	5/14	5/15	6/16	6/17	7/17
12	2	2	3	4	4/13	5/14	6/16	6/16	7/17	7/18
13	2	2	3	4	5	5/15	6/16	6/17	7/18	7/19
14	2	2	3	4	5	5/15	6/16	7/17	7/18	8/19
15	2	3	3	4	5	6/15	6/16	7/18	7/18	8/19
16	2	3	4	4	5	6	6/17	7/18	8/19	8/20
17	2	3	4	4	5	6	7/17	7/18	8/19	9/20
18	2	3	4	5	5	6	7/17	8/18	8/19	9/20
19	2	3	4	5	6	6	7/17	8/18	8/20	9/21
20	2	3	4	5	6	6	7/17	8/18	9/20	9/21

n_2 \ n_1	12	13	14	15	16	17	18	19	20
2	2	2	2	2	2	2	2	2	2
3	2	2	2	3	3	3	3	3	3
4	3	3	3	3	4	4	4	4	4
5	4	4	4	4	4	4	5	5	5
6	4/13	5	5	5	5	5	5	6	6
7	5/14	5/15	5/15	6/15	6	6	6	6	6
8	6/16	6/16	6/16	6/16	6/17	7/17	7/17	7/17	7/17
9	6/16	6/17	7/17	7/18	7/18	7/18	8/18	8/18	8/18
10	7/17	7/18	7/18	7/18	8/19	8/19	8/19	8/20	9/20
11	7/18	7/19	8/19	8/19	8/20	9/20	9/20	9/21	9/21
12	7/19	8/19	8/20	8/20	9/21	9/21	9/21	10/22	10/22
13	8/19	8/20	9/20	9/21	9/21	10/22	10/22	10/23	10/23
14	8/20	9/20	9/21	9/22	10/22	10/23	10/23	11/23	11/24
15	8/20	9/21	9/22	10/22	10/23	11/23	11/24	11/24	12/25
16	9/21	9/21	10/22	10/23	11/23	11/24	11/25	12/25	12/25
17	9/21	10/22	10/23	11/23	11/24	11/25	12/25	12/26	13/26
18	9/21	10/22	10/23	11/24	11/25	12/25	12/26	13/26	13/27
19	10/22	10/23	11/23	11/24	12/25	12/26	13/26	13/27	13/27
20	10/22	10/23	11/24	12/25	12/25	13/26	13/27	13/27	14/28

Reject H_0 if calculated number of runs is **less than or equal to** the **first** critical value **or greater than or equal to** the **second** critical value. Where only one critical value is given, reject H_0 if observed number of runs is **less than or equal to** the critical value.

C3a Critical Values of U for a One-Tailed Test at the 0.05 Significance Level or a Two-Tailed Test at the 0.1 Level

n_x \ n_y	1	2	3	4	5	6	7	8	9	10	11	12	13	14	15	16	17	18	19	20
1																			0	0
2					0	0	0	1	1	1	1	2	2	2	3	3	3	4	4	4
3					1	2	2	3	3	4	5	5	6	7	7	8	9	9	10	11
4				1	2	3	4	5	6	7	8	9	10	11	12	14	15	16	17	18
5		0	1	2	4	5	6	8	9	11	12	13	15	16	18	19	20	22	23	25
6		0	2	3	5	7	8	10	12	14	16	17	19	21	23	25	26	28	30	32
7		0	2	4	6	8	11	13	15	17	19	21	24	26	28	30	33	35	37	39
8		1	3	5	8	10	13	15	18	20	23	26	28	31	33	36	39	41	44	47
9		1	3	6	9	12	15	18	21	24	27	30	33	36	39	42	45	48	51	54
10		1	4	7	11	14	17	20	24	27	31	34	37	41	44	48	51	55	58	62
11		1	5	8	12	16	19	23	27	31	34	38	42	46	50	54	57	61	65	69
12		2	5	9	13	17	21	26	30	34	38	42	47	51	55	60	64	68	72	77
13		2	6	10	15	19	24	28	33	37	42	47	51	56	61	65	70	75	80	84
14		2	7	11	16	21	26	31	36	41	46	51	56	61	66	71	77	82	87	92
15		3	7	12	18	23	28	33	39	44	50	55	61	66	72	77	83	88	94	100
16		3	8	14	19	25	30	36	42	48	54	60	65	71	77	83	89	95	101	107
17		3	9	15	20	26	33	39	45	51	57	64	70	77	83	89	96	102	109	115
18		4	9	16	22	28	35	41	48	55	61	68	75	82	88	95	102	109	116	123
19	0	4	10	17	23	30	37	44	51	58	65	72	80	87	94	101	109	116	123	130
20	0	4	11	18	25	32	39	47	54	62	69	77	84	92	100	107	115	123	130	138

Reject H_0 if calculated value of U is **less than or equal to** critical value at chosen significance level.

C3b Critical Values of U for a One-Tailed Test at the 0.025 Significance Level or a Two-Tailed Test at the 0.05 Level

n_x \ n_y	1	2	3	4	5	6	7	8	9	10	11	12	13	14	15	16	17	18	19	20
1																				
2								0	0	0	0	1	1	1	1	1	2	2	2	2
3				0	1	1	2	2		3	3	4	4	5	5	6	6	7	7	8
4			0	1	2	3	4	4		5	6	7	8	9	10	11	11	12	13	13
5		0	1	2	3	5	6	7		8	9	11	12	13	14	15	17	18	19	20
6		1	2	3	5	6	8	10		11	13	14	16	17	19	21	22	24	25	27
7		1	3	5	6	8	10	12		14	16	18	20	22	24	26	28	30	32	34
8	0	2	4	6	8	10	13	15		17	19	22	24	26	29	31	34	36	38	41
9	0	2	4	7	10	12	15	17		20	23	26	28	31	34	37	39	42	45	48
10	0	3	5	8	11	14	17	20		23	26	29	33	36	39	42	45	48	52	55
11	0	3	6	9	13	16	19	23		26	30	33	37	40	44	47	51	55	58	62
12	1	4	7	11	14	18	22	26		29	33	37	41	45	49	53	57	61	65	69
13	1	4	8	12	16	20	24	28		33	37	41	45	50	54	59	63	67	72	76
14	1	5	9	13	17	22	26	31		36	40	45	50	55	59	64	76	74	78	83
15	1	5	10	14	19	24	29	34		39	44	49	54	59	64	70	75	80	85	90
16	1	6	11	15	21	26	31	37		42	47	53	59	64	70	75	81	86	92	98
17	2	6	11	17	22	28	34	39		45	51	57	63	67	75	81	87	93	99	105
18	2	7	12	18	24	30	36	42		48	55	61	67	74	80	86	93	99	106	112
19	2	7	13	19	25	32	38	45		52	58	65	72	78	85	92	99	106	113	119
20	2	8	13	20	27	34	41	48		55	62	69	76	83	90	98	105	112	119	127

Reject H_0 if calculated value of U is **less than or equal to** critical value at chosen significance level.

C3c Critical Values of U for a One-Tailed Test at the 0.01 Significance Level or a Two-Tailed Test at the 0.02 Level

n_x \ n_y	1	2	3	4	5	6	7	8	9	10	11	12	13	14	15	16	17	18	19	20
1																				
2										0	0	0	0	0	0		1	1		
3						0	1	1		1	1	2	2	2	3	3	4	4	4	5
4				0	1	1	2	3		3	4	5	5	6	7	7	8	9	9	10
5			0	1	2	3	4	5		6	7	8	9	10	11	12	13	14	15	16
6			1	2	3	4	6	7		8	9	11	12	13	15	16	18	19	20	22
7		0	1	3	4	6	8	9		11	12	14	16	17	19	21	23	24	26	28
8		1	2	4	6	8	10	11		13	15	17	20	22	24	26	28	30	32	34
9		1	3	5	7	9	11	14		16	18	21	23	26	28	31	33	36	38	40
10		1	3	6	8	11	13	16		19	22	24	27	30	33	36	38	41	44	47
11		1	4	7	9	12	15	18		22	25	28	31	34	37	41	44	47	50	53
12		2	5	8	11	14	17	21		24	28	31	35	38	42	46	49	53	56	60
13	0	2	5	9	12	16	20	23		27	31	35	39	43	47	51	55	59	63	67
14	0	2	6	10	13	17	22	26		30	34	38	43	47	51	56	60	65	69	73
15	0	3	7	11	15	19	24	28		33	37	42	47	51	56	61	66	70	75	80
16	0	3	7	12	16	21	26	31		36	41	46	51	56	61	66	71	76	82	87
17	0	4	8	13	18	23	28	33		38	44	49	55	60	66	71	77	82	88	93
18	0	4	9	14	19	24	30	36		41	47	53	59	65	70	76	82	88	94	100
19	1	4	9	15	20	26	32	38		44	50	56	63	69	75	82	88	94	101	107
20	1	5	10	16	22	28	34	40		47	53	60	67	73	80	87	93	100	107	114

Reject H_0 if calculated value of U is **less than or equal to** critical value at chosen significance level.

C3d Critical Values of U for a One-Tailed Test at the 0.001 Significance Level or a Two-Tailed Test at the 0.002 Level

n_x \ n_y	1	2	3	4	5	6	7	8	9	10	11	12	13	14	15	16	17	18	19	20
1																				
2																				
3																	0	0	0	0
4										0	0	0	1	1	1	2	2	3	3	3
5							0	0	1	1	2	2	3	3	4	5	5	6	7	7
6						0	1	2	2	3	4	4	5	6	7	8	9	10	11	12
7					0	1	2	3	3	5	6	7	8	9	10	11	13	14	15	16
8					0	2	3	4	5	6	8	9	11	12	14	15	17	18	20	21
9					1	2	3	5	7	8	10	12	14	15	17	19	21	23	25	26
10				0	1	3	5	6	8	10	12	14	17	19	21	23	25	27	29	32
11				0	2	4	6	8	10	12	15	17	20	22	24	27	29	32	34	37
12				0	2	4	7	9	12	14	17	20	23	25	28	31	34	37	40	42
13				1	3	5	8	11	14	17	20	23	26	29	32	35	38	42	45	48
14				1	3	6	9	12	15	19	22	25	29	32	36	39	43	46	50	54
15				1	4	7	10	14	17	21	24	28	32	36	40	43	47	51	55	59
16				2	5	8	11	15	19	23	27	31	35	39	43	48	52	56	60	65
17			0	2	5	9	13	17	21	25	29	34	38	43	47	52	57	61	66	70
18			0	3	6	10	14	18	23	27	32	37	42	46	51	56	61	66	71	76
19			0	3	7	11	15	20	25	29	34	40	45	50	55	60	66	71	77	82
20			0	3	7	12	16	21	26	32	37	42	48	54	59	65	70	76	82	88

Reject H_o if calculated value of U is **less than or equal to** critical value at chosen significance level.

Other critical values of U for use with n_x or n_y greater than 20 can be derived using the following equation:

$$U = \frac{n_x n_y}{2} - z\sqrt{\frac{n_x n_y (n_x + n_y + 1)}{12}}$$

where z is the appropriate critical value of a standard normal deviate (from Appendix C10) at the desired significance level. For example, the critical value of z for a two-tailed test at the 0.05 significance level is 1.960. The critical value of U when $n_x = 25$ and $n_y = 30$ is therefore:

$$\frac{25 \times 30}{2} - 1.960\sqrt{\frac{25 \times 30(25 + 30 + 1)}{12}} = 259$$

C4 Critical Values of Student's t

Degrees of freedom	Significance level (one-tailed)				
	0.05	0.025	0.01	0.005	0.0005
	Significance level (two-tailed)				
	0.1	0.05	0.02	0.01	0.001
1	6.31	12.71	31.82	63.66	636.62
2	2.92	4.30	6.97	9.93	31.60
3	2.35	3.18	4.54	5.84	12.92
4	2.13	2.78	3.75	4.60	8.61
5	2.01	2.57	3.37	4.03	6.86
6	1.94	2.45	3.14	3.71	5.96
7	1.89	2.37	3.00	3.50	5.41
8	1.86	2.31	2.90	3.35	5.04
9	1.83	2.26	2.82	3.25	4.78
10	1.81	2.23	2.76	3.17	4.59
11	1.80	2.20	2.72	3.11	4.44
12	1.78	2.18	2.68	3.05	4.32
13	1.77	2.16	2.65	3.01	4.22
14	1.76	2.15	2.62	2.98	4.14
15	1.75	2.13	2.60	2.95	4.07
16	1.75	2.12	2.58	2.92	4.01
17	1.74	2.11	2.57	2.90	3.97
18	1.73	2.10	2.55	2.88	3.92
19	1.73	2.09	2.54	2.86	3.88
20	1.73	2.09	2.53	2.85	3.85
21	1.72	2.08	2.52	2.83	3.82
22	1.72	2.07	2.51	2.82	3.79
23	1.71	2.07	2.50	2.81	3.77
24	1.71	2.06	2.49	2.80	3.75
25	1.71	2.06	2.49	2.79	3.73
26	1.71	2.06	2.48	2.78	3.71
27	1.70	2.05	2.47	2.77	3.69
28	1.70	2.05	2.47	2.76	3.67
29	1.70	2.05	2.46	2.76	3.66
30	1.70	2.04	2.46	2.75	3.65
40	1.68	2.02	2.42	2.70	3.55
60	1.67	2.00	2.39	2.66	3.46
120	1.66	1.98	2.36	2.62	3.37
∞	1.65	1.96	2.33	2.58	3.29

Reject H_0 if calculated value of t is **greater than** critical value at chosen significance level.

C5 Critical Values of Chi Square

Degrees of freedom	Significance level				
	0.1	0.05	0.01	0.005	0.001
1	2.71	3.84	6.64	7.88	10.83
2	4.60	5.99	9.21	10.60	13.82
3	6.25	7.82	11.34	12.84	16.27
4	7.78	9.49	13.28	14.86	18.46
5	9.24	11.07	15.09	16.75	20.52
6	10.64	12.59	16.81	18.55	22.46
7	12.02	14.07	18.48	20.28	24.32
8	13.36	15.51	20.09	21.96	26.12
9	14.68	16.92	21.67	23.59	27.88
10	15.99	18.31	23.21	25.19	29.59
11	17.28	19.68	24.72	26.76	31.26
12	18.55	21.03	26.22	28.30	32.91
13	19.81	22.36	27.69	30.82	34.53
14	21.06	23.68	29.14	31.32	36.12
15	22.31	25.00	30.58	32.80	37.70
16	23.54	26.30	32.00	34.27	39.29
17	24.77	27.59	33.41	35.72	40.75
18	25.99	28.87	34.80	37.16	42.31
19	27.20	30.14	36.19	38.58	43.82
20	28.41	31.41	37.57	40.00	45.32
21	29.62	32.67	38.93	41.40	46.80
22	30.81	33.92	40.29	42.80	48.27
23	32.01	35.17	41.64	44.18	49.73
24	33.20	36.42	42.98	45.56	51.18
25	34.38	37.65	44.31	46.93	52.62
26	35.56	35.88	45.64	48.29	54.05
27	36.74	40.11	46.96	49.65	55.48
28	37.92	41.34	48.28	50.99	56.89
29	39.09	42.56	49.59	52.34	58.30
30	40.26	43.77	50.89	53.67	59.70
40	51.81	55.76	63.69	66.77	73.40
50	63.17	67.51	76.15	79.49	86.66
60	74.40	79.08	88.38	91.95	99.61
70	85.53	90.53	100.43	104.22	112.32
80	96.58	101.88	112.33	116.32	124.84
90	107.57	113.15	124.12	128.30	137.21
100	118.50	124.34	135.81	140.17	149.45

Reject H_0 if calculated value of chi square is **greater than** the critical value at the chosen significance level.

C6 Critical Values of H for the Kruskal–Wallis Test

n_1	n_2	n_3	0.1	Significance level 0.05	0.01	0.005
2	1	1				
2	2	1				
2	2	2	4.571			
3	1	1				
3	2	1	4.286			
3	2	2	4.500	4.714	5.357	
3	3	1	4.571	5.143		
3	3	2	4.556	5.361		
3	3	3	4.622	5.600	7.200	7.200
4	1	1				
4	2	1	4.500			
4	2	2	4.056	5.208		
4	3	2	4.511	5.444	6.444	
4	3	3	4.709	5.727	6.746	
4	4	1	4.167	4.967	6.667	
4	4	2	4.555	5.455	7.036	
4	4	3	4.546	5.599	7.144	
4	4	4	4.654	5.692	7.654	
5	1	1				
5	2	1	4.200	5.000		
5	2	2	4.373	5.160	6.533	
5	3	1	4.018	4.960		
5	3	2	4.651	5.251	6.882	
5	3	3	4.533	5.649	7.079	
5	4	1	3.987	4.986	6.955	
5	4	2	4.541	5.268	7.118	
5	4	3	4.549	5.631	7.445	
5	4	4	4.619	5.618	7.760	
5	5	1	4.109	5.127	7.309	
5	5	2	4.508	5.339	7.269	
5	5	3	4.545	5.706	7.543	
5	5	4	4.523	5.643	7.791	
5	5	5	4.560	5.780	7.980	

Reject H_o if calculated value of H is **greater than or equal to** critical value at chosen significance level.

C7a Critical Values of F at the 0.1 Significance Level

Degrees of freedom for between samples variance estimate

		1	2	3	4	5	6	7	8	9
	1	39.86	49.50	53.59	55.83	57.24	58.20	58.91	59.44	59.86
	2	8.53	9.00	9.16	9.24	9.29	9.33	9.35	9.37	9.38
	3	5.54	5.46	5.39	5.34	5.31	5.28	5.27	5.25	5.24
	4	4.54	4.32	4.19	4.11	4.05	4.01	3.98	3.95	3.94
	5	4.06	3.78	3.62	3.52	3.45	3.40	3.37	3.34	3.32
	6	3.78	3.46	3.29	3.18	3.11	3.05	3.01	2.98	2.96
	7	3.59	3.26	3.07	2.96	2.88	2.83	2.78	2.75	2.72
	8	3.46	3.11	2.92	2.81	2.73	2.67	2.62	2.59	2.56
	9	3.36	3.01	2.81	2.69	2.61	2.55	2.51	2.47	2.44
	10	3.29	2.92	2.73	2.61	2.52	2.46	2.41	2.38	2.35
	11	3.23	2.86	2.66	2.54	2.45	2.39	2.34	2.30	2.27
	12	3.18	2.81	2.61	2.48	2.39	2.33	2.28	2.24	2.21
	13	3.14	2.76	2.56	2.43	2.35	2.28	2.23	2.20	2.16
	14	3.10	2.73	2.52	2.39	2.31	2.24	2.19	2.15	2.12
	15	3.07	2.70	2.49	2.36	2.27	2.21	2.16	2.12	2.09
	16	3.05	2.67	2.46	2.33	2.24	2.18	2.13	2.09	2.06
	17	3.03	2.64	2.44	2.31	2.22	2.15	2.10	2.06	2.03
	18	3.01	2.62	2.42	2.29	2.20	2.13	2.08	2.04	2.00
	19	2.99	2.61	2.40	2.27	2.18	2.11	2.06	2.02	1.98
	20	2.97	2.59	2.38	2.25	2.16	2.09	2.04	2.00	1.96
	21	2.96	2.57	2.36	2.23	2.14	2.08	2.02	1.98	1.95
	22	2.95	2.56	2.35	2.22	2.13	2.06	2.01	1.97	1.93
	23	2.94	2.55	2.34	2.21	2.11	2.05	1.99	1.95	1.92
	24	2.93	2.54	2.33	2.19	2.10	2.04	1.98	1.94	1.91
	25	2.92	2.53	2.32	2.18	2.09	2.02	1.97	1.93	1.89
	26	2.91	2.52	2.31	2.17	2.08	2.01	1.96	1.92	1.88
	27	2.90	2.51	2.30	2.17	2.07	2.00	1.95	1.91	1.87
	28	2.89	2.50	2.29	2.16	2.06	2.00	1.94	1.90	1.87
	29	2.89	2.50	2.28	2.15	2.06	1.99	1.93	1.89	1.86
	30	2.88	2.49	2.28	2.14	2.05	1.98	1.93	1.88	1.85
	40	2.84	2.44	2.23	2.09	2.00	1.93	1.87	1.83	1.79
	60	2.79	2.39	2.18	2.04	1.95	1.87	1.82	1.77	1.74
	120	2.75	2.35	2.13	1.99	1.90	1.82	1.77	1.72	1.68
	∞	2.71	2.30	2.08	1.94	1.85	1.77	1.72	1.67	1.63

Degrees of freedom for within samples variance estimate

Reject H_0 if calculated value of F is **greater than** critical value.

10	12	15	20	24	30	40	60	120	∞
60.19	60.71	61.22	61.74	62.00	62.26	62.53	62.79	63.06	63.33
9.39	9.41	9.42	9.44	9.45	9.46	9.47	9.47	9.48	9.49
5.23	5.22	5.20	5.18	5.18	5.17	5.15	5.16	5.14	5.13
3.92	3.90	3.87	3.84	3.83	3.82	3.80	3.79	3.78	3.76
3.30	3.27	3.24	3.21	3.19	3.17	3.16	3.14	3.12	3.10
2.94	2.90	2.87	2.84	2.82	2.80	2.78	2.76	2.74	2.72
2.70	2.67	2.63	2.59	2.58	2.56	2.54	2.51	2.49	2.47
2.54	2.50	2.46	2.42	2.40	2.38	2.36	2.34	2.32	2.29
2.42	2.38	2.34	2.30	2.28	2.25	2.23	2.21	2.18	2.16
2.32	2.28	2.24	2.20	2.18	2.16	2.13	2.11	2.08	2.06
2.25	2.21	2.17	2.12	2.10	2.08	2.05	2.03	2.00	1.97
2.19	2.15	2.10	2.06	2.04	2.01	1.99	1.96	1.93	1.90
2.14	2.10	2.05	2.01	1.98	1.96	1.93	1.90	1.88	1.85
2.10	2.05	2.01	1.96	1.94	1.91	1.89	1.86	1.83	1.80
2.06	2.02	1.97	1.92	1.90	1.87	1.85	1.82	1.79	1.76
2.03	1.99	1.94	1.89	1.87	1.84	1.81	1.78	1.75	1.72
2.00	1.96	1.91	1.86	1.84	1.81	1.78	1.75	1.72	1.69
1.98	1.93	1.89	1.84	1.81	1.78	1.75	1.72	1.69	1.66
1.96	1.91	1.86	1.81	1.79	1.76	1.73	1.70	1.67	1.63
1.94	1.89	1.84	1.79	1.77	1.74	1.71	1.68	1.64	1.61
1.92	1.87	1.83	1.78	1.75	1.72	1.69	1.66	1.62	1.59
1.90	1.86	1.81	1.76	1.73	1.70	1.67	1.64	1.60	1.57
1.89	1.84	1.80	1.74	1.72	1.69	1.66	1.62	1.59	1.55
1.88	1.83	1.78	1.73	1.70	1.67	1.64	1.61	1.57	1.53
1.87	1.82	1.77	1.72	1.69	1.66	1.63	1.59	1.56	1.52
1.86	1.81	1.76	1.71	1.68	1.65	1.61	1.58	1.54	1.50
1.85	1.80	1.75	1.70	1.67	1.64	1.60	1.57	1.53	1.49
1.84	1.79	1.74	1.69	1.66	1.63	1.59	1.56	1.52	1.48
1.83	1.78	1.73	1.68	1.65	1.62	1.58	1.55	1.51	1.47
1.82	1.77	1.72	1.67	1.64	1.61	1.57	1.54	1.50	1.46
1.76	1.71	1.66	1.61	1.57	1.54	1.51	1.47	1.42	1.38
1.71	1.66	1.60	1.54	1.51	1.48	1.44	1.40	1.35	1.29
1.65	1.60	1.55	1.48	1.45	1.41	1.37	1.32	1.26	1.19
1.60	1.55	1.49	1.42	1.38	1.34	1.30	1.24	1.17	1.00

Reject H_0 if calculated value of F is **greater than** critical value.

C7b Critical Values of F at the 0.05 Significance Level

Degrees of freedom for between samples variance estimate

		1	2	3	4	5	6	7	8	9
	1	161.4	199.5	215.7	224.6	230.2	234.0	236.8	238.9	240.5
	2	18.51	19.00	19.16	19.25	19.30	19.33	19.35	19.37	19.38
	3	10.13	9.55	9.28	9.12	9.01	8.94	8.89	8.85	8.81
	4	7.71	6.94	6.59	6.39	6.26	6.16	6.09	6.04	6.00
	5	6.61	5.79	5.41	5.19	5.05	4.95	4.88	4.82	4.77
	6	5.99	5.14	4.76	4.53	4.39	4.28	4.21	4.15	4.10
	7	5.59	4.74	4.35	4.12	3.97	3.87	3.79	3.73	3.68
	8	5.32	4.46	4.07	3.84	3.69	3.58	3.50	3.44	3.39
	9	5.12	4.26	3.86	3.63	3.48	3.37	3.29	3.23	3.18
	10	4.96	4.10	3.71	3.48	3.33	3.22	3.14	3.07	3.02
	11	4.84	3.98	3.59	3.36	3.20	3.09	3.01	2.95	2.90
	12	4.75	3.89	3.49	3.26	3.11	3.00	2.91	2.85	2.80
	13	4.67	3.81	3.41	3.18	3.03	2.92	2.83	2.77	2.71
	14	4.60	3.74	3.34	3.11	2.96	2.85	2.76	2.70	2.65
	15	4.54	3.68	3.29	3.06	2.90	2.79	2.71	2.64	2.59
	16	4.49	3.63	3.24	3.01	2.85	2.74	2.66	2.59	2.54
	17	4.45	3.59	3.20	2.96	2.81	2.70	2.61	2.55	2.49
	18	4.41	3.55	3.16	2.93	2.77	2.66	2.58	2.51	2.46
	19	4.38	3.52	3.13	2.90	2.74	2.63	2.54	2.48	2.42
	20	4.35	3.49	3.10	2.87	2.71	2.60	2.51	2.45	2.39
	21	4.32	3.47	3.07	2.84	2.68	2.57	2.49	2.42	2.37
	22	4.30	3.44	3.05	2.82	2.66	2.55	2.46	2.40	2.34
	23	4.28	3.42	3.03	2.80	2.64	2.53	2.44	2.37	2.32
	24	4.26	3.40	3.01	2.78	2.62	2.51	2.42	2.36	2.30
	25	4.24	3.39	2.99	2.76	2.60	2.49	2.40	2.34	2.28
	26	4.23	3.37	2.98	2.74	2.59	2.47	2.39	2.32	2.27
	27	4.21	3.35	2.96	2.73	2.57	2.46	2.37	2.31	2.25
	28	4.20	3.34	2.95	2.71	2.56	2.45	2.36	2.29	2.24
	29	4.18	3.33	2.93	2.70	2.55	2.43	2.35	2.28	2.22
	30	4.17	3.32	2.92	2.69	2.53	2.42	2.33	2.27	2.21
	40	4.08	3.23	2.84	2.61	2.45	2.34	2.25	2.18	2.12
	60	4.00	3.15	2.76	2.53	2.37	2.25	2.17	2.10	2.04
	120	3.92	3.07	2.68	2.45	2.29	2.17	2.09	2.02	1.96
	∞	3.84	3.00	2.60	2.37	2.21	2.10	2.01	1.94	1.88

Degrees of freedom for within samples variance estimate

10	12	15	20	24	30	40	60	120	∞
241.9	243.9	245.9	248.0	249.1	250.1	251.1	252.2	253.3	254.3
19.40	19.41	19.43	19.45	19.45	19.46	19.47	19.48	19.49	19.50
8.79	8.74	8.70	8.66	8.64	8.62	8.59	8.57	8.55	8.53
5.96	5.91	5.86	5.80	5.77	5.75	5.72	5.69	5.66	5.63
4.74	4.68	4.62	4.56	4.53	4.50	4.46	4.43	4.40	4.36
4.06	4.00	3.94	3.87	3.84	3.81	3.77	3.74	3.70	3.67
3.64	3.57	3.51	3.44	3.41	3.38	3.34	3.30	3.27	3.23
3.35	3.28	3.22	3.15	3.12	3.08	3.04	3.01	2.97	2.93
3.14	3.07	3.01	2.94	2.90	2.86	2.83	2.79	2.75	2.71
2.98	2.91	2.85	2.77	2.74	2.70	2.66	2.62	2.58	2.54
2.85	2.79	2.72	2.65	2.61	2.57	2.53	2.49	2.45	2.40
2.75	2.69	2.62	2.54	2.51	2.47	2.43	2.38	2.34	2.30
2.67	2.60	2.53	2.46	2.42	2.38	2.34	2.30	2.25	2.21
2.60	2.53	2.46	2.39	2.35	2.31	2.27	2.22	2.18	2.13
2.54	2.48	2.40	2.33	2.29	2.25	2.20	2.16	2.11	2.07
2.49	2.42	2.35	2.28	2.24	2.19	2.15	2.11	2.06	2.01
2.45	2.38	2.31	2.23	2.19	2.15	2.10	2.06	2.01	1.96
2.41	2.34	2.27	2.19	2.15	2.11	2.06	2.02	1.97	1.92
2.38	2.31	2.23	2.16	2.11	2.07	2.03	1.98	1.93	1.88
2.35	2.28	2.20	2.12	2.08	2.04	1.99	1.95	1.90	1.84
2.32	2.25	2.18	2.10	2.05	2.01	1.96	1.92	1.87	1.81
2.30	2.23	2.15	2.07	2.03	1.98	1.94	1.89	1.84	1.78
2.27	2.20	2.13	2.05	2.01	1.96	1.91	1.86	1.81	1.76
2.25	2.18	2.11	2.03	1.98	1.94	1.89	1.84	1.79	1.73
2.24	2.16	2.09	2.01	1.96	1.92	1.87	1.82	1.77	1.71
2.22	2.15	2.07	1.99	1.95	1.90	1.85	1.80	1.75	1.69
2.20	2.13	2.06	1.97	1.93	1.88	1.84	1.79	1.73	1.67
2.19	2.12	2.04	1.96	1.91	1.87	1.82	1.77	1.71	1.65
2.18	2.10	2.03	1.94	1.90	1.85	1.81	1.75	1.70	1.64
2.16	2.09	2.01	1.93	1.89	1.84	1.79	1.74	1.68	1.62
2.08	2.00	1.92	1.84	1.79	1.74	1.69	1.64	1.58	1.51
1.99	1.92	1.84	1.75	1.70	1.65	1.59	1.53	1.47	1.39
1.91	1.83	1.75	1.66	1.61	1.55	1.50	1.43	1.35	1.25
1.83	1.75	1.67	1.57	1.52	1.46	1.39	1.32	1.22	1.00

Reject H_0 if calculated value of F is **greater than** critical value.

C7c Critical Values of F at the 0.01 Significance Level

Degrees of freedom for between samples variance estimate

		1	2	3	4	5	6	7	8	9
	1	4052	4999.5	5403	5625	5764	5859	5928	5982	6022
	2	98.50	99.00	99.17	99.25	99.30	99.33	99.36	99.37	99.39
	3	34.12	30.82	29.46	28.71	28.24	27.91	27.67	27.49	27.35
	4	21.20	18.00	16.69	15.98	15.52	15.21	14.98	14.80	14.66
	5	16.26	13.27	12.06	11.39	10.97	10.67	10.46	10.29	10.16
	6	13.75	10.92	9.78	9.15	8.75	8.47	8.26	8.10	7.98
	7	12.25	9.55	8.45	7.85	7.46	7.19	6.99	6.84	6.72
	8	11.26	8.65	7.59	7.01	6.63	6.37	6.18	6.03	5.91
	9	10.56	8.02	6.99	6.42	6.06	5.80	5.61	5.47	5.35
	10	10.04	7.56	6.55	5.99	5.64	5.39	5.20	5.06	4.94
	11	9.65	7.21	6.22	5.67	5.32	5.07	4.89	4.74	4.63
	12	9.33	6.93	5.95	5.41	5.06	4.82	4.64	4.50	4.39
	13	9.07	6.70	5.74	5.21	4.86	4.62	4.44	4.30	4.19
Degrees of freedom for within samples variance estimate	14	8.86	6.51	5.56	5.04	4.69	4.46	4.28	4.14	4.03
	15	8.68	6.36	5.42	4.89	4.56	4.32	4.14	4.00	3.89
	16	8.53	6.23	5.29	4.77	4.44	4.20	4.03	3.89	3.78
	17	8.40	6.11	5.18	4.67	4.34	4.10	3.93	3.79	3.68
	18	8.29	6.01	5.09	4.58	4.25	4.01	3.84	3.71	3.60
	19	8.18	5.93	5.01	4.50	4.17	3.94	3.77	3.63	3.52
	20	8.10	5.85	4.94	4.43	4.10	3.87	3.70	3.56	3.46
	21	8.02	5.78	4.87	4.37	4.04	3.81	3.64	3.51	3.40
	22	7.95	5.72	4.82	4.31	3.99	3.76	3.59	3.45	3.35
	23	7.88	5.66	4.76	4.26	3.94	3.71	3.54	3.41	3.30
	24	7.82	5.61	4.72	4.22	3.90	3.67	3.50	3.36	3.26
	25	7.77	5.57	4.68	4.18	3.85	3.63	3.46	3.32	3.22
	26	7.72	5.53	4.64	4.14	3.82	3.59	3.42	3.29	3.18
	27	7.68	5.49	4.60	4.11	3.78	3.56	3.39	3.26	3.15
	28	7.64	5.45	4.57	4.07	3.75	3.53	3.36	3.23	3.12
	29	7.60	5.42	4.54	4.04	3.73	3.50	3.33	3.20	3.09
	30	7.56	5.39	4.51	4.02	3.70	3.47	3.30	3.17	3.07
	40	7.31	5.18	4.31	3.83	3.51	3.29	3.12	2.99	2.89
	60	7.08	4.98	4.13	3.65	3.34	3.12	2.95	2.82	2.72
	120	6.85	4.79	3.95	3.48	3.17	2.96	2.79	2.66	2.56
	∞	6.63	4.61	3.78	3.32	3.02	2.80	2.64	2.51	2.41

10	12	15	20	24	30	40	60	120	∞
6056	6106	6157	6209	6235	6261	6287	6313	6339	6366
99.40	99.42	99.43	99.45	99.46	99.47	99.47	99.48	99.49	99.50
27.23	27.05	26.87	26.69	26.60	26.50	26.41	26.32	26.22	26.13
14.55	14.37	14.20	14.02	13.93	13.84	13.75	13.65	13.56	13.46
10.05	9.89	9.72	9.55	9.47	9.38	9.29	9.20	9.11	9.02
7.87	7.72	7.56	7.40	7.31	7.23	7.14	7.06	6.97	6.88
6.62	6.47	6.31	6.16	6.07	5.99	5.91	5.82	5.74	5.65
5.81	5.67	5.52	5.36	5.28	5.20	5.12	5.03	4.95	4.86
5.26	5.11	4.96	4.81	4.73	4.65	4.57	4.48	4.40	4.31
4.85	4.71	4.56	4.41	4.33	4.25	4.17	4.08	4.00	3.91
4.54	4.40	4.25	4.10	4.02	3.94	3.86	3.78	3.69	3.60
4.30	4.16	4.01	3.86	3.78	3.70	3.62	3.54	3.45	3.36
4.10	3.96	3.82	3.66	3.59	3.51	3.43	3.34	3.25	3.17
3.94	3.80	3.66	3.51	3.43	3.35	3.27	3.18	3.09	3.00
3.80	3.67	3.52	3.37	3.29	3.21	3.13	3.05	2.96	2.87
3.69	3.55	3.41	3.26	3.18	3.10	3.02	2.93	2.84	2.75
3.59	3.46	3.31	3.16	3.08	3.00	2.92	2.83	2.75	2.65
3.51	3.37	3.23	3.08	3.00	2.92	2.84	2.75	2.66	2.57
3.43	3.30	3.15	3.00	2.92	2.84	2.76	2.67	2.58	2.49
3.37	3.23	3.09	2.94	2.86	2.78	2.69	2.61	2.52	2.42
3.31	3.17	3.03	2.88	2.80	2.72	2.64	2.55	2.46	2.36
3.26	3.12	2.98	2.83	2.75	2.67	2.58	2.50	2.40	2.31
3.21	3.07	2.93	2.78	2.70	2.62	2.54	2.45	2.35	2.26
3.17	3.03	2.89	2.74	2.66	2.58	2.49	2.40	2.31	2.21
3.13	2.99	2.85	2.70	2.62	2.54	2.45	2.36	2.27	2.17
3.09	2.96	2.81	2.66	2.58	2.50	2.42	2.33	2.23	2.13
3.06	2.93	2.78	2.63	2.55	2.47	2.38	2.29	2.20	2.10
3.03	2.90	2.75	2.60	2.52	2.44	2.35	2.26	2.17	2.06
3.00	2.87	2.73	2.57	2.49	2.41	2.33	2.23	2.14	2.03
2.98	2.84	2.70	2.55	2.47	2.39	2.30	2.21	2.11	2.01
2.80	2.66	2.52	2.37	2.29	2.20	2.11	2.02	1.92	1.80
2.63	2.50	2.35	2.20	2.12	2.03	1.94	1.84	1.73	1.60
2.47	2.34	2.19	2.03	1.95	1.86	1.76	1.66	1.53	1.38
2.32	2.18	2.04	1.88	1.79	1.70	1.59	1.47	1.32	1.00

Reject H_0 if calculated value of F is **greater than** critical value.

C7d Critical Values of F at the 0.005 Significance Level

Degrees of freedom for between samples variance estimate

		1	2	3	4	5	6	7	8	9
	1	16211	20000	21615	22500	23056	23437	23715	23925	24091
	2	198.5	199.0	199.2	199.2	199.3	199.3	199.4	199.4	199.4
	3	55.55	49.80	47.47	46.19	45.39	44.84	44.43	44.13	43.88
		31.33	26.28	24.26	23.15	22.46	21.97	21.62	21.35	21.14
	5	22.78	18.31	16.53	15.56	14.94	14.51	14.20	13.96	13.77
	6	18.63	14.54	12.92	12.03	11.46	11.07	10.79	10.57	10.39
	7	16.24	12.40	10.88	10.05	9.52	9.16	8.89	8.68	8.51
	8	14.69	11.04	9.60	8.81	8.30	7.95	7.69	7.50	7.34
	9	13.61	10.11	8.72	7.96	7.47	7.13	6.88	6.69	6.54
	10	12.83	9.43	8.08	7.34	6.87	6.54	6.30	6.12	5.97
	11	12.23	8.91	7.60	6.88	6.42	6.10	5.86	5.68	5.54
	12	11.75	8.51	7.23	6.52	6.07	5.76	5.52	5.35	5.20
	13	11.37	8.19	6.93	6.23	5.79	5.48	5.25	5.08	4.94
	14	11.06	7.92	6.68	6.00	5.56	5.26	5.03	4.86	4.72
	15	10.80	7.70	6.48	5.80	5.37	5.07	4.85	4.67	4.54
	16	10.58	7.51	6.30	5.64	5.21	4.91	4.69	4.52	4.38
	17	10.38	7.35	6.16	5.50	5.07	4.78	4.56	4.39	4.25
	18	10.22	7.21	6.03	5.37	4.96	4.66	4.44	4.28	4.14
	19	10.07	7.09	5.92	5.27	4.85	4.56	4.34	4.18	4.04
	20	9.94	6.99	5.82	5.17	4.76	4.47	4.26	4.09	3.96
	21	9.83	6.89	5.73	5.09	4.68	4.39	4.18	4.01	3.88
	22	9.73	6.81	5.65	5.02	4.61	4.32	4.11	3.94	3.81
	23	9.63	6.73	5.58	4.95	4.54	4.26	4.05	3.88	3.75
	24	9.55	6.66	5.52	4.89	4.49	4.20	3.99	3.83	3.69
	25	9.48	6.60	5.46	4.84	4.43	4.15	3.94	3.78	3.64
	26	9.41	6.54	5.41	4.79	4.38	4.10	3.89	3.73	3.60
	27	9.34	6.49	5.36	4.74	4.34	4.06	3.85	3.69	3.56
	28	9.28	6.44	5.32	4.70	4.30	4.02	3.81	3.65	3.52
	29	9.23	6.40	5.28	4.66	4.26	3.98	3.77	3.61	3.48
	30	9.18	6.35	5.24	4.62	4.23	3.95	3.74	3.58	3.45
	40	8.83	6.07	4.98	4.37	3.99	3.71	3.51	3.35	3.22
	60	8.49	5.79	4.73	4.14	3.76	3.49	3.29	3.13	3.01
	120	8.18	5.54	4.50	3.92	3.55	3.28	3.09	2.93	2.81
	∞	7.88	5.30	4.28	3.72	3.35	3.09	2.90	2.74	2.62

Degrees of freedom for within samples variance estimate

10	12	15	20	24	30	40	60	120	∞
24224	24426	24630	24836	24940	25044	25148	25253	25359	25465
199.4	199.4	199.4	199.4	199.5	199.5	199.5	199.5	199.5	199.5
43.69	43.39	43.08	42.78	42.62	42.47	42.31	42.15	41.99	41.83
20.97	20.70	20.44	20.17	20.03	19.89	19.75	19.61	19.47	19.32
13.62	13.38	13.15	12.90	12.78	12.66	12.53	12.40	12.27	12.14
10.25	10.03	9.81	9.59	9.47	9.36	9.24	9.12	9.00	8.88
8.38	8.18	7.97	7.75	7.65	7.53	7.42	7.31	7.19	7.08
7.21	7.01	6.81	6.61	6.50	6.40	6.29	6.18	6.06	5.95
6.42	6.23	6.03	5.83	5.73	5.62	5.52	5.41	5.30	5.19
5.85	5.66	5.47	5.27	5.17	5.07	4.97	4.86	4.75	4.64
5.42	5.24	5.05	4.86	4.76	4.65	4.55	4.44	4.34	4.23
5.09	4.91	4.72	4.53	4.43	4.33	4.23	4.12	4.01	3.90
4.82	4.64	4.46	4.27	4.17	4.07	3.97	3.87	3.76	3.65
4.60	4.43	4.25	4.06	3.96	3.86	3.76	3.66	3.55	3.44
4.42	4.25	4.07	3.88	3.79	3.69	3.58	3.48	3.37	3.26
4.27	4.10	3.92	3.73	3.64	3.54	3.44	3.33	3.22	3.11
4.14	3.97	3.79	3.61	3.51	3.41	3.31	3.21	3.10	2.98
4.03	3.86	3.68	3.50	3.40	3.30	3.20	3.10	2.99	2.87
3.93	3.76	3.59	3.40	3.31	3.21	3.11	3.00	2.89	2.78
3.85	3.68	3.50	3.32	3.22	3.12	3.02	2.92	2.81	2.69
3.77	3.60	3.43	3.24	3.15	3.05	2.95	2.84	2.73	2.61
3.70	3.54	3.36	3.18	3.08	2.98	2.88	2.77	2.66	2.55
3.64	3.47	3.30	3.12	3.02	2.92	2.82	2.71	2.60	2.48
3.59	3.42	3.25	3.06	2.97	2.87	2.77	2.66	2.55	2.43
3.54	3.37	3.20	3.01	2.92	2.82	2.72	2.61	2.50	2.38
3.49	3.33	3.15	2.97	2.87	2.77	2.67	2.56	2.45	2.33
3.45	3.28	3.11	2.93	2.83	2.73	2.63	2.52	2.41	2.25
3.41	3.25	3.07	2.89	2.79	2.69	2.59	2.48	2.37	2.29
3.38	3.21	3.04	2.86	2.76	2.66	2.56	2.45	2.33	2.24
3.34	3.18	3.01	2.82	2.73	2.63	2.52	2.42	2.30	2.18
3.12	2.95	2.78	2.60	2.50	2.40	2.30	2.18	2.06	1.93
2.90	2.74	2.57	2.39	2.29	2.19	2.08	1.96	1.83	1.69
2.71	2.54	2.37	2.19	2.09	1.98	1.87	1.75	1.61	1.43
2.52	2.36	2.19	2.00	1.90	1.79	1.67	1.53	1.36	1.00

Reject H_0 if calculated value of F is **greater than** critical value.

C7e Critical Values of F at the 0.001 Significance Level

Degrees of freedom for between samples variance estimate

		1	2	3	4	5	6	7	8	9
	1	405300	500000	540400	562500	576400	585900	592900	598100	602300
	2	998.5	999.0	999.2	999.2	999.3	999.3	999.4	999.4	999.4
	3	167.0	148.5	141.1	137.1	134.6	132.8	131.6	130.6	129.9
	4	74.14	61.25	56.18	53.44	51.71	50.53	49.66	49.00	48.47
	5	47.18	37.12	33.20	31.09	29.75	28.84	28.16	27.64	27.24
	6	35.51	27.00	23.70	21.92	20.81	20.03	19.46	19.03	18.69
	7	29.25	21.69	18.77	17.19	16.21	15.52	15.02	14.63	14.33
	8	25.42	18.49	15.83	14.39	13.49	12.86	12.40	12.04	11.77
	9	22.86	16.39	13.90	12.56	11.71	11.13	10.70	10.37	10.11
	10	21.04	14.91	12.55	11.28	10.48	9.92	9.52	9.20	8.96
	11	19.69	13.81	11.56	10.35	9.58	9.05	8.66	8.35	8.12
	12	18.64	12.97	10.80	9.63	8.89	8.38	8.00	7.71	7.48
	13	17.81	12.31	10.21	9.07	8.35	7.86	7.49	7.21	6.98
	14	17.14	11.78	9.73	8.62	7.92	7.43	7.08	6.80	6.58
	15	16.59	11.34	9.34	8.25	7.57	7.09	6.74	6.47	6.26
	16	16.12	10.97	9.00	7.94	7.27	6.81	6.46	6.19	5.98
	17	15.72	10.66	8.73	7.68	7.02	6.56	6.22	5.96	5.75
	18	15.38	10.39	8.49	7.46	6.81	6.35	6.02	5.76	5.56
	19	15.08	10.16	8.28	7.26	6.62	6.18	5.85	5.59	5.39
	20	14.82	9.95	8.10	7.10	6.46	6.02	5.69	5.44	5.24
	21	14.59	9.77	7.94	6.95	6.32	5.88	5.56	5.31	5.11
	22	14.38	9.61	7.80	6.81	6.19	5.76	5.44	5.19	4.99
	23	14.19	9.47	7.67	6.69	6.08	5.65	5.33	5.09	4.89
	24	14.03	9.34	7.55	6.59	5.98	5.55	5.23	4.99	4.80
	25	13.88	9.22	7.45	6.49	5.88	5.46	5.15	4.91	4.71
	26	13.74	9.12	7.36	6.41	5.80	5.38	5.07	4.83	4.64
	27	13.61	9.02	7.27	6.33	5.73	5.31	5.00	4.76	4.57
	28	13.50	8.93	7.19	6.25	5.66	5.24	4.93	4.69	4.50
	29	13.39	8.85	7.12	6.19	5.59	5.18	4.87	4.64	4.45
	30	13.29	8.77	7.05	6.12	5.53	5.12	4.82	4.58	4.39
	40	12.61	8.25	6.60	5.70	5.13	4.73	4.44	4.21	4.02
	60	11.97	7.76	6.17	5.31	4.76	4.37	4.09	3.87	3.69
	120	11.38	7.32	5.79	4.95	4.42	4.04	3.77	3.55	3.38
	∞	10.83	6.91	5.42	4.62	4.10	3.74	3.47	3.27	3.10

Degrees of freedom for within samples variance estimate

10	12	15	20	24	30	40	60	120	∞
605600	610700	615800	620900	623500	626100	628700	631300	634000	636600
999.4	999.4	999.4	999.4	999.5	999.5	999.5	999.5	999.5	999.5
129.2	128.3	127.4	126.4	125.9	125.4	125.0	124.5	124.0	123.5
48.05	47.41	46.76	46.10	45.77	45.43	45.09	44.75	44.40	44.05
26.92	26.42	25.91	25.39	25.14	24.87	24.60	24.33	24.06	23.79
18.41	17.99	17.56	17.12	16.89	16.67	16.44	16.21	15.99	15.75
14.08	13.71	13.32	12.93	12.73	12.53	12.33	12.12	11.91	11.70
11.54	11.19	10.84	10.48	10.30	10.11	9.92	9.73	9.53	9.33
9.89	9.57	9.24	8.90	8.72	8.55	8.37	8.19	8.00	7.81
8.75	8.45	8.13	7.80	7.64	7.47	7.30	7.12	6.94	6.76
7.92	7.63	7.32	7.01	6.85	6.68	6.52	6.35	6.17	6.00
7.29	7.00	6.71	6.40	6.25	6.09	5.93	5.76	5.59	5.42
6.80	6.52	6.23	5.93	5.78	5.63	5.47	5.30	5.14	4.97
6.40	6.13	5.85	5.56	5.41	5.25	5.10	4.94	4.77	4.60
6.08	5.81	5.54	5.25	5.10	4.95	4.80	4.64	4.47	4.31
5.81	5.55	5.27	4.99	4.85	4.70	4.54	4.39	4.23	4.06
5.58	5.32	5.05	4.78	4.63	4.48	4.33	4.18	4.02	3.85
5.39	5.13	4.87	4.59	4.45	4.30	4.15	4.00	3.84	3.67
5.22	4.97	4.70	4.43	4.29	4.14	3.99	3.84	3.68	3.51
5.08	4.82	4.56	4.29	4.15	4.00	3.86	3.70	3.54	3.38
4.95	4.70	4.44	4.17	4.03	3.88	3.74	3.58	3.42	3.26
4.83	4.58	4.33	4.06	3.92	3.78	3.63	3.48	3.32	3.15
4.73	4.48	4.23	3.96	3.82	3.68	3.53	3.38	3.22	3.05
4.64	4.39	4.14	3.87	3.74	3.59	3.45	3.29	3.14	2.97
4.56	4.31	4.06	3.79	3.66	3.52	3.37	3.22	3.06	2.89
4.48	4.24	3.99	3.72	3.59	3.44	3.30	3.15	2.99	2.82
4.41	4.17	3.92	3.66	3.52	3.38	3.23	3.08	2.92	2.75
4.35	4.11	3.86	3.60	3.46	3.32	3.18	3.02	2.86	2.69
4.29	4.05	3.80	3.54	3.41	3.27	3.12	2.97	2.81	2.64
4.24	4.00	3.75	3.49	3.36	3.22	3.07	2.92	2.76	2.59
3.87	3.64	3.40	3.15	3.01	2.87	2.73	2.57	2.41	2.23
3.54	3.31	3.08	2.83	2.69	2.55	2.41	2.25	2.08	1.89
3.24	3.02	2.78	2.53	2.40	2.26	2.11	1.95	1.76	1.54
2.96	2.74	2.51	2.27	2.13	1.99	1.84	1.66	1.45	1.00

Reject H_0 if calculated value of F is **greater than** critical value

C8 Critical Values of Pearson's Product-Moment Correlation Coefficient r

Degrees of freedom	Significance level (one-tailed)			
	0.05	0.025	0.01	0.005
	Significance level (two-tailed)			
	0.1	0.05	0.02	0.01
1	0.9877	0.9969	0.9995	0.9999
2	0.900	0.950	0.980	0.990
3	0.805	0.878	0.934	0.959
4	0.729	0.811	0.882	0.917
5	0.669	0.755	0.833	0.875
6	0.622	0.707	0.789	0.834
7	0.582	0.666	0.750	0.798
8	0.549	0.632	0.716	0.765
9	0.521	0.602	0.685	0.735
10	0.497	0.576	0.658	0.708
11	0.476	0.553	0.634	0.684
12	0.458	0.532	0.612	0.661
13	0.441	0.514	0.592	0.641
14	0.426	0.497	0.574	0.623
15	0.412	0.482	0.558	0.606
16	0.400	0.468	0.543	0.590
17	0.389	0.456	0.529	0.575
18	0.378	0.444	0.516	0.561
19	0.369	0.433	0.503	0.549
20	0.360	0.423	0.492	0.537
25	0.323	0.381	0.445	0.487
30	0.296	0.349	0.409	0.449
35	0.275	0.325	0.381	0.418
40	0.257	0.304	0.358	0.393
45	0.243	0.288	0.338	0.372
50	0.231	0.273	0.322	0.354
60	0.211	0.250	0.295	0.325
70	0.195	0.232	0.274	0.302
80	0.183	0.217	0.257	0.283
90	0.173	0.205	0.242	0.267
100	0.164	0.195	0.230	0.254

Reject H_0 if calculated value of r is **greater than** critical value at chosen significance level (in absolute terms).

C9 Critical Values of Spearman's Rank Correlation Coefficient r_s

Degrees of freedom	Significance level (one-tailed)			
	0.05	0.025	0.01	0.005
	Significance level (two-tailed)			
	0.1	0.05	0.02	0.01
4	1.000			
5	0.900	1.000	1.000	
6	0.829	0.886	0.943	1.000
7	0.714	0.786	0.893	0.929
8	0.643	0.738	0.833	0.881
9	0.600	0.683	0.783	0.833
10	0.564	0.648	0.745	0.794
11	0.523	0.623	0.736	0.818
12	0.497	0.591	0.703	0.780
13	0.475	0.566	0.673	0.745
14	0.457	0.545	0.646	0.716
15	0.441	0.525	0.623	0.689
16	0.425	0.507	0.601	0.666
17	0.412	0.490	0.582	0.645
18	0.399	0.476	0.564	0.625
19	0.388	0.462	0.549	0.608
20	0.377	0.450	0.534	0.591
21	0.368	0.438	0.521	0.576
22	0.359	0.428	0.508	0.562
23	0.351	0.418	0.496	0.549
24	0.343	0.409	0.485	0.537
25	0.336	0.400	0.475	0.526
26	0.329	0.392	0.465	0.515
27	0.323	0.385	0.456	0.505
28	0.317	0.377	0.448	0.496
29	0.311	0.370	0.440	0.487
30	0.305	0.364	0.432	0.478
35	0.282	0.336	0.399	0.442
40	0.263	0.314	0.373	0.413
45	0.248	0.296	0.351	0.388
50	0.235	0.280	0.332	0.368
55	0.224	0.267	0.317	0.351
60	0.214	0.255	0.303	0.335
65	0.206	0.245	0.291	0.322
70	0.198	0.236	0.280	0.310
75	0.191	0.228	0.271	0.300
80	0.185	0.221	0.262	0.290
85	0.180	0.214	0.254	0.281
90	0.174	0.208	0.247	0.273
95	0.170	0.202	0.240	0.266
100	0.165	0.197	0.234	0.259

Reject H_0 if calculated value of r_s is **greater than** the critical value at the chosen significance level (in absolute terms).

For degrees of freedom greater than 30 other critical values can be found from the following relationship:

$$r_s = z\sqrt{1/(n-1)}$$

where r_s is the critical value of r_s, n is the number of individuals in the data set (the degrees of freedom), and z is the appropriate critical value of a standard normal deviate (from Appendix C10). For a two-tailed test at the 0.01 level the appropriate value of z is 2.576, so the critical value of r_s with 72 degrees of freedom is:

$$2.576\sqrt{1/(72-1)} = 2.576\sqrt{0.014}$$
$$= 2.576 \times 0.119$$
$$= 0.306$$

C10 Critical Values of a Standard Normal Deviate z

| | Significance level (one-tailed) | | | | |
	0.1	0.05	0.01	0.005	0.001
z	1.282	1.645	2.326	2.576	3.090
−z	−1.282	−1.645	−2.326	−2.576	−3.090

| | Significance level (two-tailed) | | | | |
	0.1	0.05	0.01	0.005	0.001
z	1.645	1.960	2.576	2.813	3.291
−z	−1.645	−1.960	−2.576	−2.813	−3.291

C11a Critical Values of the Nearest-Neighbour Index, R (One-Tailed)

n	Clustered pattern					Dispersed pattern				
	0.1	0.05	0.01	0.005	0.001	0.1	0.05	0.01	0.005	0.001
2	0.527	0.392	0.140	0.048		1.473	1.608	1.860	1.952	2.139
3	0.614	0.504	0.298	0.223	0.071	1.386	1.497	1.702	1.777	1.930
4	0.666	0.570	0.392	0.327	0.195	1.335	1.430	1.608	1.673	1.805
5	0.701	0.616	0.456	0.398	0.280	1.299	1.385	1.544	1.602	1.720
6	0.727	0.649	0.504	0.451	0.343	1.273	1.351	1.497	1.550	1.657
7	0.747	0.675	0.540	0.491	0.392	1.253	1.325	1.460	1.509	1.609
8	0.764	0.696	0.570	0.524	0.431	1.237	1.304	1.430	1.476	1.569
9	0.777	0.713	0.595	0.551	0.463	1.223	1.287	1.406	1.449	1.537
10	0.789	0.728	0.615	0.574	0.491	1.212	1.272	1.385	1.426	1.509
11	0.798	0.741	0.633	0.594	0.515	1.202	1.259	1.367	1.406	1.486
12	0.807	0.752	0.649	0.612	0.535	1.193	1.248	1.351	1.389	1.465
13	0.815	0.762	0.663	0.627	0.554	1.186	1.239	1.337	1.373	1.447
14	0.821	0.770	0.675	0.640	0.570	1.179	1.230	1.325	1.360	1.430
15	0.827	0.778	0.686	0.653	0.584	1.173	1.222	1.314	1.348	1.416
16	0.833	0.785	0.696	0.664	0.598	1.167	1.215	1.304	1.337	1.403
17	0.838	0.792	0.705	0.674	0.610	1.162	1.209	1.295	1.327	1.391
18	0.842	0.797	0.713	0.683	0.621	1.158	1.203	1.287	1.317	1.380
19	0.847	0.803	0.721	0.691	0.631	1.154	1.197	1.279	1.309	1.369
20	0.850	0.808	0.728	0.699	0.640	1.150	1.192	1.272	1.301	1.360
21	0.854	0.812	0.735	0.706	0.649	1.146	1.188	1.266	1.294	1.351
22	0.857	0.817	0.741	0.713	0.657	1.143	1.183	1.259	1.287	1.343
23	0.861	0.821	0.746	0.719	0.664	1.140	1.179	1.254	1.281	1.336
24	0.864	0.825	0.752	0.725	0.671	1.137	1.176	1.248	1.275	1.329
25	0.866	0.828	0.757	0.731	0.678	1.134	1.172	1.243	1.269	1.322
26	0.869	0.831	0.762	0.736	0.684	1.131	1.169	1.239	1.264	1.316
27	0.871	0.835	0.766	0.741	0.690	1.129	1.166	1.234	1.259	1.310
28	0.874	0.838	0.770	0.746	0.696	1.127	1.163	1.230	1.254	1.304
29	0.876	0.840	0.774	0.750	0.701	1.124	1.160	1.226	1.250	1.299
30	0.878	0.843	0.778	0.754	0.706	1.122	1.157	1.222	1.246	1.294
31	0.880	0.846	0.782	0.758	0.711	1.120	1.155	1.219	1.242	1.289
32	0.882	0.848	0.785	0.762	0.715	1.118	1.152	1.215	1.238	1.285
33	0.884	0.850	0.788	0.766	0.720	1.117	1.150	1.212	1.234	1.280
34	0.885	0.853	0.791	0.769	0.724	1.115	1.148	1.209	1.231	1.276
35	0.887	0.855	0.794	0.773	0.728	1.113	1.145	1.206	1.228	1.272
36	0.889	0.857	0.797	0.776	0.732	1.112	1.143	1.203	1.224	1.268
37	0.890	0.859	0.800	0.779	0.735	1.110	1.141	1.200	1.221	1.265
38	0.892	0.861	0.803	0.782	0.739	1.109	1.140	1.197	1.218	1.261
39	0.893	0.862	0.805	0.785	0.742	1.107	1.138	1.195	1.216	1.258
40	0.894	0.864	0.808	0.787	0.746	1.106	1.136	1.192	1.213	1.255
41	0.896	0.866	0.810	0.790	0.749	1.105	1.134	1.190	1.210	1.252
42	0.897	0.867	0.812	0.792	0.752	1.103	1.133	1.188	1.208	1.249
43	0.898	0.869	0.815	0.795	0.755	1.102	1.131	1.186	1.205	1.246
44	0.899	0.870	0.817	0.797	0.757	1.101	1.130	1.183	1.203	1.243

n	Clustered pattern					Dispersed pattern				
	0.1	0.05	0.01	0.005	0.001	0.1	0.05	0.01	0.005	0.001
45	0.900	0.872	0.819	0.799	0.760	1.100	1.128	1.181	1.201	1.240
46	0.901	0.873	0.821	0.802	0.763	1.099	1.127	1.179	1.199	1.237
47	0.903	0.875	0.823	0.804	0.765	1.098	1.126	1.178	1.196	1.235
48	0.904	0.876	0.825	0.806	0.768	1.097	1.124	1.176	1.194	1.232
49	0.905	0.877	0.826	0.808	0.770	1.096	1.123	1.174	1.192	1.230
50	0.905	0.878	0.828	0.810	0.772	1.095	1.122	1.172	1.190	1.228
55	0.910	0.884	0.836	0.819	0.783	1.090	1.116	1.164	1.182	1.217
60	0.914	0.889	0.843	0.826	0.792	1.086	1.111	1.157	1.174	1.208
65	0.917	0.893	0.849	0.833	0.800	1.083	1.107	1.151	1.167	1.200
70	0.920	0.897	0.855	0.839	0.808	1.080	1.103	1.145	1.161	1.193
75	0.923	0.901	0.860	0.845	0.814	1.077	1.099	1.141	1.156	1.186
80	0.925	0.904	0.864	0.850	0.820	1.075	1.096	1.136	1.151	1.180
85	0.928	0.907	0.868	0.854	0.825	1.073	1.093	1.132	1.146	1.175
90	0.930	0.909	0.872	0.858	0.830	1.071	1.091	1.128	1.142	1.170
95	0.931	0.912	0.875	0.862	0.835	1.069	1.088	1.125	1.138	1.165
100	0.933	0.914	0.878	0.865	0.839	1.067	1.086	1.122	1.135	1.161
200	0.953	0.939	0.914	0.905	0.886	1.047	1.061	1.086	1.095	1.114
300	0.961	0.950	0.930	0.922	0.907	1.039	1.050	1.070	1.078	1.093
400	0.967	0.957	0.939	0.933	0.920	1.034	1.043	1.061	1.067	1.081
500	0.970	0.962	0.946	0.940	0.928	1.030	1.039	1.054	1.060	1.072

To test for clustering: reject H_0 if calculated value of R is **less than** critical value at chosen significance level

To test for dispersion: reject H_0 if calculated value of R is **greater than** critical value at chosen significance level

C11b Critical Values of the Nearest-Neighbour Index, R (Two-Tailed)

n	Significance level									
	0.1		0.05		0.01		0.005		0.001	
2	0.392	1.608	0.276	1.725	0.048	1.952		2.040		2.217
3	0.504	1.497	0.409	1.592	0.223	1.778	0.151	1.849	0.007	1.993
4	0.570	1.430	0.488	1.512	0.327	1.673	0.265	1.735	0.140	1.860
5	0.616	1.385	0.542	1.458	0.398	1.602	0.343	1.658	0.231	1.769
6	0.649	1.351	0.582	1.418	0.450	1.550	0.400	1.600	0.298	1.702
7	0.675	1.325	0.613	1.387	0.491	1.509	0.444	1.556	0.350	1.650
8	0.696	1.304	0.638	1.362	0.524	1.476	0.480	1.520	0.392	1.608
9	0.713	1.287	0.659	1.342	0.551	1.449	0.510	1.490	0.427	1.574
10	0.728	1.272	0.676	1.324	0.574	1.426	0.535	1.465	0.456	1.544
11	0.741	1.259	0.691	1.309	0.594	1.406	0.557	1.443	0.481	1.519
12	0.752	1.248	0.704	1.296	0.611	1.389	0.576	1.425	0.504	1.497
13	0.762	1.239	0.716	1.284	0.627	1.374	0.592	1.408	0.523	1.477
14	0.770	1.230	0.726	1.274	0.640	1.360	0.607	1.393	0.540	1.460
15	0.778	1.222	0.736	1.265	0.652	1.348	0.620	1.380	0.556	1.444
16	0.785	1.215	0.744	1.256	0.663	1.337	0.632	1.368	0.570	1.430
17	0.792	1.209	0.752	1.249	0.674	1.327	0.643	1.357	0.583	1.417
18	0.797	1.203	0.759	1.242	0.683	1.317	0.654	1.347	0.595	1.406
19	0.803	1.197	0.765	1.235	0.691	1.309	0.663	1.337	0.605	1.395

	Significance level									
	0.1		0.05		0.01		0.005		0.001	
20	0.808	1.192	0.771	1.229	0.699	1.301	0.671	1.329	0.615	1.385
21	0.812	1.188	0.777	1.224	0.706	1.294	0.679	1.321	0.625	1.375
22	0.817	1.183	0.782	1.219	0.713	1.287	0.687	1.314	0.633	1.367
23	0.821	1.179	0.786	1.214	0.719	1.281	0.693	1.307	0.641	1.359
24	0.825	1.176	0.791	1.209	0.725	1.275	0.700	1.300	0.649	1.351
25	0.828	1.172	0.795	1.205	0.731	1.269	0.706	1.294	0.656	1.344
26	0.831	1.169	0.799	1.201	0.736	1.264	0.712	1.288	0.663	1.337
27	0.835	1.166	0.803	1.197	0.741	1.259	0.717	1.283	0.669	1.331
28	0.838	1.163	0.806	1.194	0.746	1.255	0.722	1.278	0.675	1.325
29	0.840	1.160	0.810	1.190	0.750	1.250	0.727	1.273	0.681	1.320
30	0.843	1.157	0.813	1.187	0.754	1.246	0.732	1.269	0.686	1.314
31	0.846	1.155	0.816	1.184	0.758	1.242	0.736	1.264	0.691	1.309
32	0.848	1.152	0.819	1.181	0.762	1.238	0.740	1.260	0.696	1.304
33	0.850	1.150	0.822	1.178	0.766	1.235	0.744	1.256	0.701	1.300
34	0.853	1.148	0.824	1.176	0.769	1.231	0.748	1.252	0.705	1.295
35	0.855	1.145	0.827	1.173	0.772	1.228	0.752	1.249	0.709	1.291
36	0.857	1.143	0.829	1.171	0.776	1.225	0.755	1.245	0.713	1.287
37	0.859	1.141	0.832	1.169	0.779	1.221	0.758	1.242	0.717	1.283
38	0.861	1.140	0.834	1.166	0.782	1.219	0.762	1.239	0.721	1.279
39	0.862	1.138	0.836	1.164	0.784	1.216	0.765	1.236	0.725	1.276
40	0.864	1.136	0.838	1.162	0.787	1.213	0.768	1.233	0.728	1.272
41	0.866	1.134	0.840	1.160	0.790	1.210	0.770	1.230	0.731	1.269
42	0.867	1.133	0.842	1.158	0.792	1.208	0.773	1.227	0.735	1.266
43	0.869	1.131	0.844	1.156	0.795	1.205	0.776	1.224	0.738	1.262
44	0.870	1.130	0.846	1.155	0.797	1.203	0.778	1.222	0.741	1.259
45	0.872	1.128	0.847	1.153	0.799	1.201	0.781	1.219	0.744	1.257
46	0.873	1.127	0.849	1.151	0.802	1.199	0.783	1.217	0.746	1.254
47	0.875	1.126	0.851	1.150	0.804	1.197	0.786	1.215	0.749	1.251
48	0.876	1.124	0.852	1.148	0.806	1.194	0.788	1.212	0.752	1.248
49	0.877	1.123	0.854	1.146	0.808	1.192	0.790	1.210	0.754	1.246
50	0.878	1.122	0.855	1.145	0.810	1.191	0.792	1.208	0.757	1.243
55	0.884	1.116	0.862	1.138	0.819	1.182	0.802	1.198	0.768	1.232
60	0.889	1.111	0.868	1.132	0.826	1.174	0.810	1.190	0.778	1.222
65	0.893	1.107	0.873	1.127	0.833	1.167	0.818	1.182	0.787	1.213
70	0.897	1.103	0.878	1.123	0.839	1.161	0.824	1.176	0.794	1.206
75	0.901	1.099	0.882	1.118	0.845	1.156	0.830	1.170	0.801	1.199
80	0.904	1.096	0.886	1.115	0.850	1.151	0.836	1.164	0.808	1.192
85	0.907	1.093	0.889	1.111	0.854	1.146	0.841	1.160	0.814	1.187
90	0.909	1.091	0.892	1.108	0.858	1.142	0.845	1.155	0.819	1.181
95	0.912	1.088	0.895	1.105	0.862	1.138	0.849	1.151	0.824	1.177
100	0.914	1.086	0.898	1.103	0.865	1.135	0.853	1.147	0.828	1.172
200	0.939	1.061	0.928	1.073	0.905	1.095	0.896	1.104	0.878	1.122
300	0.950	1.050	0.941	1.059	0.922	1.078	0.915	1.085	0.901	1.099
400	0.957	1.043	0.949	1.051	0.933	1.067	0.927	1.074	0.914	1.086
500	0.962	1.039	0.954	1.046	0.940	1.060	0.934	1.066	0.923	1.077

Reject H_0 if calculated value of R is **less than lower** critical value or **greater than higher** critical value at chosen significance level.

APPENDIX D

Random Numbers

D1 Random Numbers

80 30	23 64	67 96	21 33	36 90	03 91	69 33	90 13	34 43	02 19
61 29	89 61	32 08	12 62	26 08	42 00	31 73	31 30	30 61	34 11
23 33	61 01	02 21	11 81	51 32	36 10	23 74	50 31	90 11	73 52
94 21	32 92	93 50	72 67	23 20	74 59	30 30	43 66	75 32	27 97
87 61	92 69	01 60	28 79	74 76	86 06	39 29	73 85	03 27	50 57
37 56	19 18	03 42	86 03	85 74	44 81	86 45	71 16	13 52	35 56
64 86	66 31	55 04	88 40	10 30	84 38	06 13	58 83	62 04	63 52
22 69	22 69	58 45	49 23	09 81	98 84	05 04	75 99	27 70	72 79
23 22	14 22	64 90	10 26	74 23	53 91	27 73	78 19	92 43	68 10
42 38	59 64	72 96	46 57	89 67	22 81	94 56	69 84	18 31	06 39
17 18	01 34	10 98	37 48	93 86	88 59	69 53	78 86	37 26	85 48
39 45	69 53	94 89	58 97	29 33	29 19	50 94	80 57	31 99	38 91
43 18	11 42	56 19	48 44	45 02	84 29	01 78	65 77	76 84	88 85
59 44	06 45	68 55	16 65	66 13	38 00	95 76	50 67	67 65	18 83
01 50	34 32	38 00	37 57	47 82	66 59	19 50	87 14	35 59	79 47
79 14	60 35	47 95	90 71	31 03	85 37	38 70	34 16	64 55	66 49
01 56	63 68	80 26	14 97	23 88	59 22	82 39	70 83	48 34	46 48
25 76	18 71	29 25	15 51	92 96	01 01	28 18	03 35	11 10	27 84
23 52	10 83	45 06	49 85	35 45	84 08	81 13	52 57	21 23	67 02
91 64	08 64	25 74	16 10	97 31	10 27	24 48	89 06	42 81	29 10
80 86	07 27	26 70	08 65	85 20	31 23	28 99	39 63	32 03	71 91
31 71	37 60	95 60	94 95	54 45	27 97	03 67	30 54	86 04	12 41
05 83	50 36	09 04	39 15	66 55	80 36	39 71	24 10	62 22	21 53
98 70	02 90	30 63	62 59	26 04	97 20	00 91	28 80	40 23	09 91
82 79	35 45	64 53	93 24	86 55	48 72	18 57	05 79	20 09	31 46
37 52	49 55	40 65	27 61	08 59	91 23	26 18	95 04	98 20	99 52
48 16	69 65	69 02	08 83	08 83	68 37	00 96	13 59	12 16	17 93
50 43	06 59	56 53	30 61	50 21	29 06	49 60	90 38	31 43	19 25
89 31	62 79	45 73	71 72	77 11	28 80	72 35	75 77	24 72	98 43
63 29	90 61	86 39	07 38	38 85	77 06	10 23	30 84	07 95	30 76

71 68	93 94	08 72	36 27	85 89	40 59	83 37	93 85	73 97	84 05
05 06	96 63	58 24	05 95	56 64	77 53	85 64	15 95	93 91	59 03
03 35	58 95	46 44	25 70	31 66	01 05	44 44	62 91	36 31	45 04
13 04	57 67	74 77	53 35	93 51	82 83	27 38	63 16	04 48	75 23
49 96	43 94	56 04	02 79	55 78	01 44	72 26	85 54	01 81	32 82
24 36	24 08	44 77	57 07	54 41	04 56	09 44	30 58	25 45	37 56
55 19	97 20	01 11	47 45	79 79	06 72	12 81	86 97	54 09	06 53
02 28	54 60	28 35	32 94	36 74	51 63	96 90	04 13	30 43	10 14
90 50	13 78	22 20	37 56	97 95	49 95	91 15	52 73	12 93	78 94
33 71	32 43	29 58	47 38	39 96	67 51	64 47	49 91	64 58	93 07
70 58	28 49	54 32	97 70	27 81	64 69	71 52	02 56	61 37	04 58
09 68	96 1U	57 78	85 00	89 81	98 30	19 40	76 28	62 99	99 83
19 36	60 85	35 04	12 87	83 88	66 54	32 00	30 20	05 30	42 63
04 75	44 49	64 26	51 46	80 50	53 91	00 55	67 36	68 66	08 29
79 83	32 39	46 77	56 83	42 21	60 03	14 47	07 01	66 85	49 22
80 99	42 43	05 58	54 41	98 05	54 39	34 42	97 47	38 35	59 40
48 83	64 99	86 94	48 78	79 20	62 23	56 45	92 65	56 36	83 02
28 45	35 85	22 20	13 01	73 96	70 05	84 50	68 59	96 58	16 63
52 07	63 15	82 30	66 23	14 26	66 61	17 80	41 97	40 27	24 80
39 14	52 18	35 87	48 55	48 81	03 11	26 99	03 80	08 86	50 42

D2 Random Numbers from a Normal Distribution

1.54	−2.46	1.16	0.86	0.72	0.79	0.35	0.05	−0.08	−0.02
−0.45	−0.74	−0.87	3.15	−1.23	−1.52	2.35	2.31	−2.00	−2.28
1.52	1.48	−2.76	0.92	0.70	0.66	0.48	0.11	−0.10	−0.14
−0.32	−0.68	−0.89	3.02	−1.10	−1.46	2.31	2.18	−1.87	−2.23
1.48	1.35	−2.63	0.97	0.67	0.54	0.60	0.16	−0.14	−0.26
−0.20	−0.63	−0.92	−1.05	2.96	−1.41	−1.70	2.15	2.11	−2.18
−2.46	1.33	1.29	−2.93	0.73	0.51	0.48	0.30	−0.07	−0.29
−0.32	−0.50	−0.86	−1.07	2.82	−1.28	−1.64	2.12	1.99	−2.05
1.60	1.29	1.16	−2.80	0.78	0.48	0.35	0.42	−0.02	−0.32
−0.45	3.61	−0.81	−1.10	−1.23	2.76	−1.59	−1.87	1.96	1.92
−2.35	1.36	1.14	1.10	−3.10	0.54	0.33	0.29	0.11	−0.26
−0.47	3.48	−0.68	−1.04	−1.25	2.63	−1.46	−1.82	1.93	1.79
−2.22	1.41	1.10	0.97	−2.98	0.59	0.30	0.17	0.23	−0.21
−0.50	−0.63	3.41	−1.00	−1.28	−1.41	2.56	−1.77	−2.05	1.77
1.73	−2.53	1.17	0.95	0.91	−3.28	0.36	0.14	0.10	−0.07
−0.44	−0.65	3.28	−0.86	−1.22	−1.43	2.43	−1.64	−1.99	1.74
1.60	−2.40	1.22	0.92	0.79	0.85	0.41	0.11	−0.02	0.05
−0.39	−0.68	−0.81	3.22	−1.18	−1.46	2.41	2.37	−1.95	−2.23
1.58	1.54	−2.70	0.98	0.76	0.72	0.54	0.17	−0.05	−0.08
−0.26	−0.62	−0.83	3.08	−1.04	−1.40	2.38	2.24	−1.81	−2.17
1.55	1.41	−2.57	1.03	0.73	0.60	0.66	0.22	−0.08	−0.20
−0.14	−0.57	−0.86	−0.99	3.02	−1.35	−1.64	2.21	2.18	−2.12
−2.40	1.39	1.35	−2.88	0.79	0.57	0.54	0.36	−0.01	−0.23
−0.27	−0.44	−0.80	−1.02	2.89	−1.22	−1.58	2.18	2.05	−1.99

−2.34	1.36	1.22	−2.75	0.84	0.54	0.41	0.48	0.04	−0.26
−0.39	3.68	−0.76	−1.05	−1.17	2.82	−1.53	−1.82	2.02	1.98
−2.30	1.42	1.20	1.16	−3.05	0.60	0.39	0.35	0.17	−0.20
−0.41	3.54	−0.62	−0.98	−1.20	2.69	−1.40	−1.76	1.99	1.86
−2.17	1.47	1.17	1.03	−2.92	0.65	0.36	0.23	0.29	−0.15
−0.44	−0.57	3.48	−0.94	−1.23	−1.35	2.63	−1.71	−1.99	1.83
1.79	−2.47	1.23	1.01	0.97	−3.22	0.42	0.20	0.16	−0.01
−0.38	−0.59	3.34	−0.80	−1.16	−1.37	2.50	−1.58	−1.94	1.80
1.66	−2.34	1.28	0.98	0.85	−3.09	0.47	0.17	0.04	0.11
−0.33	−0.62	−0.75	3.28	−1.12	−1.40	2.47	2.43	−1.89	−2.17
1.64	1.60	−2.65	1.04	0.82	0.78	0.61	0.23	0.02	−0.02
−0.20	−0.56	−0.78	3.15	−0.98	−1.34	2.44	2.30	−1.76	−2.11
1.61	1.47	−2.52	1.09	0.79	0.66	0.72	0.28	−0.02	−0.14
−0.08	−0.52	−0.81	−0.93	3.08	−1.30	−1.58	2.28	2.24	−2.06
−2.35	1.45	1.41	−2.82	0.85	0.63	0.60	0.42	0.05	−0.17
−0.21	−0.38	−0.75	−0.96	2.95	−1.16	−1.52	2.25	2.11	−1.93
−2.29	1.42	1.28	−2.69	0.90	0.60	0.47	0.54	0.10	−0.20
−0.33	−0.26	−0.70	−0.99	−1.11	2.89	−1.48	−1.76	2.08	2.05
−2.24	1.48	1.26	1.22	−2.99	0.66	0.45	0.41	0.23	−0.14
−0.35	3.61	−0.56	−0.93	−1.14	2.76	−1.34	−1.70	2.05	1.92
−2.11	1.53	1.23	1.09	−2.86	0.71	0.42	0.29	0.35	−0.09
−0.38	−0.51	3.54	−0.88	−1.17	−1.29	2.69	−1.65	−1.94	1.89
1.85	−2.42	1.29	1.07	1.03	−3.16	0.48	0.26	0.22	0.05
−0.32	−0.54	3.41	−0.74	−1.11	−1.32	2.56	−1.52	−1.88	1.86
1.73	−2.29	1.34	1.04	0.91	−3.04	0.53	0.23	0.10	0.17
−0.27	−0.57	−0.69	3.34	−1.06	−1.35	−1.47	2.50	−1.83	−2.11

Data Matrix

Individuals	A	B	C	D	E	F	G	H	I	J	K	L	M
							Variables						
Basildon	129330	39	20.38	8.3	***	****	****	0.77	1.06	0	658	12	0
Birkenhead	137852	35	12.70	6.9	5.5	8.14	1.96	0.49	0.68	0	864	30	0
Birmingham	1014670	2	12.70	7.8	10.7	2.34	0.00	0.49	0.64	1	891	163	1
Blackpool	151860	33	19.56	11.9	5.5	3.96	0.00	0.46	0.45	0	916	10	0
Bolton	154199	31	14.14	7.5	8.0	3.12	0.00	0.43	0.53	0	1259	99	0
Bournemouth	153869	32	23.85	14.1	10.0	2.76	3.62	0.61	0.97	0	976	40	0
Bradford	294177	11	13.67	8.9	11.9	6.78	0.00	0.39	0.41	1	1111	134	1
Brighton	161351	30	20.33	11.4	9.9	3.69	0.00	0.47	0.61	1	809	10	1
Bristol	426657	7	14.13	7.9	10.0	10.60	28.40	0.61	0.99	1	950	61	1
Cardiff	279111	14	13.36	9.9	6.8	2.57	37.73	0.58	0.85	1	1135	62	1
Coventry	335238	9	9.46	6.4	6.5	3.09	4.41	0.58	0.74	0	854	104	1
Derby	219582	19	13.71	5.1	7.1	1.19	0.00	0.55	0.59	1	839	48	0
Dudley	185581	23	10.24	7.4	4.3	3.49	0.00	0.64	0.72	0	837	227	0
Gateshead	103261	50	12.78	6.0	10.1	3.56	0.00	0.44	0.51	0	771	155	0
Huddersfield	131190	37	13.95	10.1	7.2	1.58	0.00	0.45	0.54	0	1229	235	0
Hull	285970	12	12.20	7.2	6.6	2.20	47.59	0.37	0.29	1	786	2	1
Ipswich	123312	41	13.38	7.6	6.0	4.20	67.76	0.59	0.59	1	652	57	0
Leeds	496009	6	13.65	9.3	9.1	2.94	5.88	0.41	0.45	1	750	23	1
Leicester	284208	13	13.72	8.8	9.2	1.10	0.00	0.47	0.61	1	708	76	1
Liverpool	610113	3	14.06	7.5	8.1	1.10	26.94	0.37	0.45	1	953	39	1
London	7452346	1	12.97	14.7	9.2	5.15	10.17	0.56	0.92	1	746	8	1
Luton	161405	29	9.42	6.9	4.3	1.87	18.60	0.62	0.74	0	802	101	0
Manchester	543650	4	14.05	6.9	13.0	3.05	16.06	0.36	0.41	1	1011	36	1
Newcastle	222209	18	15.75	7.7	14.4	3.46	0.00	0.34	0.34	1	800	48	1
Newport	112286	43	11.85	7.7	6.2	3.83	8.11	0.57	0.66	0	1150	23	0
Northampton	126642	40	15.24	9.0	7.2	2.54	1.12	0.58	0.73	0	734	58	0
Norwich	122083	42	14.91	7.2	7.7	2.49	15.06	0.54	0.66	0	715	28	1
Nottingham	300630	10	12.57	7.3	11.5	2.16	0.00	0.43	0.42	1	789	49	1
Oldham	105913	46	13.97	6.8	11.2	2.29	0.00	0.34	0.27	0	1307	167	0
Oxford	108805	45	13.79	7.9	10.5	3.88	0.00	0.58	0.91	1	750	63	1
Plymouth	239452	17	14.70	5.7	9.0	1.28	0.00	0.57	0.59	1	1147	27	0
Portsmouth	197431	22	18.08	7.2	11.0	3.83	0.00	0.48	0.59	1	771	2	0
Reading	132939	36	12.56	9.3	7.0	0.33	20.78	0.57	0.80	1	782	45	1

Individuals	Variables												
	A	B	C	D	E	F	G	H	I	J	K	L	M
Salford	130976	38	13.06	5.9	9.1	4.27	0.22	0.29	0.26	0	1017	22	1
Sheffield	520327	5	13.95	8.6	7.0	3.65	35.84	0.45	0.53	1	968	131	1
Solihull	107095	47	9.62	21.1	2.0	0.85	0.00	1.02	2.56	0	937	129	0
Southampton	215118	21	11.99	7.4	9.8	3.19	13.61	0.57	0.75	1	894	20	1
Southend	162770	28	20.15	15.5	7.2	2.67	0.00	0.58	0.90	0	630	37	0
St Helens	104341	49	12.94	4.1	7.7	2.63	0.00	0.41	0.29	0	1019	81	0
Stockport	139644	34	13.68	9.2	8.0	0.58	0.00	0.50	0.60	0	1013	85	0
Stoke	265258	16	11.31	5.5	6.8	3.73	6.58	0.46	0.41	1	922	118	0
Sunderland	216885	20	11.99	6.9	6.9	2.46	3.25	0.37	0.27	0	774	40	0
Swansea	173413	25	12.98	8.3	3.8	3.41	0.00	0.60	0.79	1	1283	10	1
Teesside	396230	8	9.97	5.5	7.7	2.77	0.00	0.47	0.47	1	759	19	0
Torbay	109257	44	22.15	15.2	5.9	2.88	0.00	0.60	0.88	0	1059	8	0
Walsall	184734	24	10.72	8.2	7.7	2.22	0.00	0.54	0.69	0	914	163	0
Warley	163567	27	13.27	5.6	3.8	1.26	0.00	0.47	0.44	0	921	210	0
West Bromwich	166593	26	10.51	6.2	5.9	2.12	0.00	0.49	0.45	0	817	109	0
Wolverhampton	269112	15	10.37	8.1	5.4	2.49	0.00	0.57	0.79	1	866	131	0
York	104782	48	14.51	7.1	7.5	2.29	0.00	0.43	0.44	0	720	9	1

(asterisks indicate data not available)

Key to variables:

A Population (*Census of England and Wales*, 1971)

B Rank size

C Percentage of population over 65 years (*Census of England and Wales*, 1971)

D Percentage of working males in social classes I and II (*Census of England and Wales*, 1971)

E Children in care of Local Authority as a percentage of all children under 18 (Institute of Municipal Treasurers, *Children's Service Statistics*, 1970–71)

F Local Authority expenditure (£) on child centres and clinics per child under 5 (Institute of Municipal Treasurers, *Children's Service Statistics*, 1970–71)

G Local Authority expenditure (£) on family planning per 1000 people (Institute of Municipal Treasurers, *Welfare Service Statistics*, 1970–71)

H Cars per household (*Census of England and Wales*, 1971)

I Percentage of households with three or more cars (*Census of England and Wales*, 1971)

J Local radio station 1976 (1 = yes, 0 = no)

K Rainfall total in millimetres 1966 (*British Rainfall*, 1966)

L Altitude of rainfall station in metres O.D. (*British Rainfall*, 1966)

M University (1 = yes, 0 = no)

Notes for Programmers

One potential problem that I am aware of concerns the data checking routine that forms part of every program. If data checking is invoked the user is shown each data value in turn and told to press RETURN if the value is correct or type in the correct value if it is not. The relevant section of program is:

```
K=0
PRINT"Press RETURN if value is correct";
PRINT" - otherwise type correct value"
FOR I=N1 TO N2
K=K+1
PRINT K;" ";X(I)
INPUT A$
IF A$<>"" X(I)=VAL(A$)
NEXT I
```

On some machines (for example the Commodore PET and CBM series) pressing RETURN in response to an INPUT command will bring the program to an abrupt halt! This problem can be overcome by substituting the following section of program:

```
K=0
PRINT"Type * if value is correct";
PRINT" - otherwise type correct value"
FOR I=N1 TO N2
K=K+1
PRINT K;" ";X(I)
INPUT A$
IF A$<>"*" X(I)=VAL(A$)
NEXT I
```

When checking data the user would then have to type * and press RETURN for each correct value. Any other character could be substituted for * in the routine. It would be convenient to use a character on a key that is near (but not too near!) the RETURN key and one that does not need the SHIFT key to be held down.

Mistakes when typing in data can be aggravated by oversensitive keyboards. Many machines have an auto-repeat on all keys, which means that characters get repeated if the key is held down too long. It may be possible, and it is certainly desirable, to switch this facility off. On the BBCb computer, for example, this can be done by typing *FX11,0 before running a program. Alternatively this command can be included in each program by inserting the following line at the beginning:

```
1 *FX11,0
```

Be careful not to overwrite an existing line!

There is also the danger with certain of the programs and on certain machines that the results may be so numerous that they disappear off the top of the screen before you have time to read them and write them down. If you have a printer connected to your computer you should be able to get all the output printed out. On a BBCb this can be done by typing the command VDU2 before running the program, or by inserting the following line in the programs:

```
2 VDU2
```

If you do not have a printer you may be able to switch on 'paging'. On a BBCb this can be done by typing the command VDU14 before running the program, or by inserting the following line:

```
3 VDU14
```

When this has been done the computer pauses once the screen is full and waits until the SHIFT key is pressed before continuing.

If your computer has the ability to work in different screen 'modes' you may find it helpful to use the mode that gives the greatest line length and the maximum number of lines per 'screenful'. On a BBCb this is mode 0, which gives 32 lines of 80 characters each. The machine can be put into this mode by typing the command MODE 0 before running the program. Alternatively, the following line can be inserted into each program:

```
4 MODE 0
```

Lower case letters have been used in various captions that are output by the programs. This is simply in the interests of clarity. If your computer does not have lower case characters the programs will work just as well if you use upper case characters throughout.

Index